A Laboratory Manual of
Food Analysis

AF657291

A Laboratory Manual of Food Analysis

Shalini Sehgal
Associate Professor
Department of Food Technology
Bhaskaracharya College of Applied Sciences
University of Delhi
Delhi

TECHSAR™
NEW DELHI

Published and Printed by
Techsar Pvt. Ltd.
4435-36/7, Ansari Road, Daryaganj,
New Delhi–110 002 (India)
E-mail: tech_info@techsarworld.com
Website: www.techsarworld.com

ISBN 978-93-84588-84-7

© 2016 I.K. International Pvt. Ltd.

Reprint 2025

All rights reserved. No part of this publication may be reproduced, stored in a retrieval system, or transmitted in any form or any means: electronic, mechanical, photocopying, recording, or otherwise, without the prior written permission from the publisher.

Preface

The need for chemical analysis of food has increased in recent years. Growing emphasis in the industry on the nutritional value of foods, and increased regulations to list nutritional contents on packages require: analysis of dietary content. This needs not only accurate analysis of organic components such as vitamins, but also inorganic nutrients such as minerals, and especially salt content. In food processing, the addition of supplements to increase the appearance or performance of food products must also be monitored. This latter need requires ongoing analysis to support the food processing operation.

Thus, food products are analyzed for a variety of reasons, e.g., compliance with legal and labeling requirements, assessment of product quality, determination of nutritive value, detection of adulteration, research and development.

This book covers the basic principles of analytical procedures and techniques commonly used to provide information about the chemical composition, structure and physical properties of food materials. The aim is to give students the experience in performing food analysis experiments, analyzing data and reporting their findings.

Prior knowledge and clear concepts about an experiment help in it being performed quickly, efficiently and safely. So, an attempt has been made to impart information about safety aspects, recording experiments in a proper format and correct representation of data along with review questions for the students.

Shalini Sehgal

Acknowledgements

The work on this manual has been an exciting, sometimes challenging, but always an interesting experience. It has been made possible by many people - chemists who have toiled from ages giving chemistry its charm which allures me till date, my parents who guided me and my friends who have encouraged me as always. I am also thankful to my dear students who have given their insights in making this manual more user friendly. Finally, I wish to thank my family for their continuous support and encouragement especially Ridhu, Aryan and Reyan for bearing with me.

Shalini Sehgal

Contents

1
Proximate Analysis of Food

Food analysis is the discipline that deals with the development, application and study of analytical procedures for characterizing the properties of foods and their constituents. These analytical procedures are used to provide information about a wide variety of different characteristics of foods, including their composition, structure, physicochemical properties and sensory attributes. This information is critical to our rational understanding of the factors that determine the properties of foods, as well as to our ability to economically produce foods that are consistently safe, nutritious and desirable and for consumers to make informed choices about their diet.

Analysis of the properties of a food material depends on the successful completion of a number of different steps: planning (identifying the most appropriate analytical procedure), sample selection, sample preparation, performance of analytical procedure, statistical analysis of measurements, and data reporting.

Many of the methods in use today for the analysis of foods are procedures based on a system introduced initially about 100 years ago by two German scientists— Henneberg and Stohmann, for the analysis of animal feedstuffs and described as the *Proximate Analysis of Foods.* This scheme of analysis involves the estimation of the main components of a food using procedures that allow a reasonably rapid and acceptable measurement of various food fractions without the need for sophisticated equipment or chemicals. The description of these food fractions, as shown in Table 1.1, remains basically the same today as in the original scheme. At present, various alternative terminologies have been introduced which, along with modifications to the analytical methods used earlier, more accurately represent the food fractions being investigated.

Although many of the basic principles of the analytical procedures remain fundamentally the same as in the original system, major advances have taken place, particularly in the use of automated equipment and of sophisticated analytical instruments enabling many of the analyses to be performed more rapidly and with a greater degree of precision.

Increased awareness and knowledge about the nutritional and functional properties of various food constituents has resulted in a greater body of information being required of particular foods, e.g., the degree of saturation of constituent fats, the levels of individual minerals and vitamins and the amounts of minor constituents such as trace elements.

Table 1.1 Proximate analysis of foods

Original terminology	*Alternative terminology*
Moisture	Loss on drying
Ash	Mineral elements
Crude fat	Fat Ether extract
Crude protein	Protein
Nitrogen-free extractives	Carbohydrates Available carbohydrates
Crude fiber	Unavailable carbohydrates Fiber Neutral detergent fiber Dietary fiber Non-starch polysaccharides

Consequently, the modern analysis of a food generally requires many more estimations to be performed than in the original scheme of analysis and this, in turn, involves the use of a wide range of different techniques and principles.

The quantitative analysis of food and food products may broadly be divided into **proximate analysis** and **ultimate analysis**. The former gives useful information, particularly from the nutritional and biochemical point of view, while the latter refers to the determination of a particular element or a compound present in the material.

The proximate analysis includes determination of the percentages of the moisture, ash, acidity, crude fat or ether extractives, protein, sugars and the crude fiber. Their sum total subtracted from 100 represents primarily the amount of carbohydrates other than sugars, but includes starch, pectin, gums, etc. In addition, all the errors in the determinations of all the other constituents are also reflected in the calculation of the carbohydrates other than sugars.

2
Sampling

Foods and food products are variable in composition, e.g., plant foods are more variable in composition than flesh foods. Processing methods cause additional changes in composition.

In sampling foods and food products, sufficient material must be taken to compensate for the variability involved. It is also advisable to analyze replicate samples considering the variability and the requirements of certain analytical techniques. The amount of material to be selected for analysis can be estimated statistically when the extent of variability of the individual sample is available. When repeated chemical analyses are done, it is advisable to make a preliminary determination of the variability of the sample. The number of individual samples to be selected may be found from the following expression:

$$n = C\sqrt{N}$$

where n = number of individual samples to be selected; C = a factor which represents the degree of accuracy desired in the sample; and N = lot size (where the extent of variability is not known. It is advisable to select at least ten times the amount to be taken as a sample for analysis).

Since most foods are relatively heterogeneous in their nature, it is important to ensure that, prior to compositional analysis, samples of the food taken for analysis are truly representative of the product to be analyzed. Sampling procedures vary from food to food and ISO standards have been set out for various foodstuffs. In general, dry foods should be brought to a powder by means of a mechanical grinder, moist solid foods should be homogenised by using equipment such as a domestic food processor, and fluid foods should be emulsified using blenders. Once prepared, food samples should be transferred as quickly as possible to dry glass or rigid plastic containers and sealed to avoid moisture loss or gain, and then clearly labeled and stored in a cool environment.

Prior to each analysis, a representative sample of the food material must be carefully prepared. The method of sampling is related to the nature of the food and should follow the guidelines indicated below:

1. Dry foods

Normally, the food should be passed through a grinder and then mixed in a mortar. Hard foods such as chocolate should be grated.

2. *Moist foods*

Foods such as meat and fish products and vegetables should be treated in the same way as dry foods and the process repeated at least once before transferring the mixture to a closed container and storing it under refrigeration.

3. *Wet foods*

Foods such as pickles, sauces and canned products are often best treated in a high speed blender. Care must be taken to avoid fat separation with emulsions such as salad cream and cream soups.

4. *Fats and oils*

These should be warmed and filtered. The temperature must not be excessive to avoid any possible losses of antioxidants present.

5. *Fatty emulsions*

Foods such as butter or margarine should be warmed to 35°C in a screw-capped jar and shaken.

REQUIREMENT FOR ANALYTICAL METHODS

The choice of method(s) used for the analysis of foods is dependent on a number of factors and relates to the following features:

1. Precision
2. Reproducibility
3. Accuracy
4. Simplicity of operation
5. Economy
6. Speed
7. Sensitivity
8. Specificity
9. Safety
10. Official approval.

QUALITY OF DATA

A number of steps and precautions may be undertaken to avoid many of the experimental errors and to improve the reliability of data produced. These include the following:

1. **Quality of glassware:** The glassware (pipettes, burettes, volumetric flasks, cuvettes, etc.) and equipment being used for

the analysis should be of a quality appropriate to the degree of precision required.

2. **Handling and cleanliness of equipment:** Glassware and equipment should be handled in the correct manner; for example, volumetric flasks, which are calibrated to specified temperatures, should not be heated. Thorough cleaning of glassware is also important for obtaining meaningful data. This may be achieved using cleaning reagents such as chromic acid or a mixture of concentrated sulfuric and nitric acids. followed by efficient rinsing first with tap water and finally with distilled water. Excessive use of detergents should be avoided.
3. **Blank analyses:** In order to ensure that background interferences from materials used in an analysis are not occurring, the analysis of a reagent blank should be carried out wherever possible. This blank should contain all the reagents used in the test sample but exclude the sample itself. Values obtained in the blank analysis should then be subtracted from those obtained with the sample being analyzed.
4. **Replication:** As many replicates as possible shall be performed in order to minimise the effects of random errors.
5. **Recovery experiments:** To measure the efficiency with which a food component, such as an additive, is being determined, samples of a food should be 'spiked' by the addition of a known amount of the component. These samples should then be analyzed to determine the percentage recovery of the added component.
6. **Reference samples:** The validity of an analytical procedure may be estimated by carrying out analyses on foods of known composition. Such standard food samples are available commercially and are an invaluable measure of the effectiveness of methods such as in the estimation of dietary fiber.
7. **Collaborative testing:** By collaboration with a number of laboratories, the results obtained by a particular laboratory may be compared with those being achieved by others using the same method. This allows the detection of any routine errors within anyone laboratory where the results are consistently different from those of other participants in the scheme.
8. **Confirmatory analysis:** The results obtained by any particular method being used should be compared against those obtained by a reference method chosen from one recognised internationally and published by bodies such as BIS, FSSAI and ISO. This allows a measure to be made for the validity of the method being used for routine purposes.

3
Safety Rules for the Laboratory

1. Follow all instructions carefully. If you don't understand what you are expected to do, ask your instructor before proceeding.
2. Be properly prepared to do the experiment. Read the written procedures in advance and understand what you are going to do.
3. Conduct yourself in a responsible manner whenever you are in the laboratory; do not touch any equipment or chemicals until you are instructed to do so.
4. Eating and drinking is not permitted in the laboratory
5. Wear a lab apron or coat while working.
6. Keep your hands away from our eyes, mouth and face when using chemicals.
7. Clean and return all equipment when instructed to do so by your instructor.
8. Follow instructions for handling chemicals. Do not taste, touch, or smell any chemicals unless told to do so. Dispose of all chemicals by following your instructor's instructions.
9. Follow instructions for the handling and dilution of acids and bases. Never add water to any concentrated acid, especially sulfuric acid. The acid must be slowly added to the water with stirring to allow the dissipation of heat involved in the dilution.
10. Make sure your hands are dry before removing an electric plug from a socket. Report any damaged electrical equipment, including, frayed wires and loose connections.
11. Be very careful when using a gas burner. Keep hair, clothing, and your hands away from an open flame. Never point the open end of a test tube that is being heated at yourself or anyone else.
12. Turn off your Bunsen burner or other heat source whenever you are not using it. Never let it operate unattended
13. Report all accidents, injuries, and close-calls to your instructor or T.A.
14. Clean all spills immediately. This includes water.
15. Never take chemicals, supplies, or equipment out of the laboratory without the knowledge and consent of the instructor.
16. Wash off chemicals splashed or spilled on your skin or body immediately

17. Pipetting by mouth should not be performed. Use pipette pumps or similar equipment. When fitting the pipette to pipette filler, hold the pipette as close as possible to the point of entry of the filler to avoid the risk of breaking the pipette and consequent injury.
18. Read labels on reagent bottles carefully and ensure the contents are exactly what are called for in the procedure. Notice the following:

19. To avoid contamination of initial reagents, do not insert droppers, pipettes or spatulas into reagent bottles. Dispense some reagent into a clean beaker and dispense from there. Do not return unused chemicals to the reagent bottle. After using a reagent bottle, replace the lid.
20. Keep the balance and the area around it clean. Do not place chemicals directly on the balance pan or dispense them into a container on the pan. Do not weigh an object while it is hot.
21. All work involving fumes or generation of aerosols must be carried out in an appropriate fume hood or bio-safety cabinet.
22. Due to its fragility and potential for serious injuries, all glassware should be handled carefully.
23. Do not use glassware that is chipped or cracked.
24. Any glass apparatus that is going to be heated or cooled must be made of borosilicate ("Pyrex") glass. "Soda" glass is normally only used for storage or volumetric glassware. For applications involving heating above 250°C, silica glassware should be used.
25. When lifting or moving hot glass beakers or flasks, always use tongs.
26. Clean your lab benches, put away all equipment and reagents, and wash your hands at the end of each weighing.

4

Preparation of Primary and Secondary Solutions

4.1 INTRODUCTION

A solution is a homogenous mixture of two or more substances. The components of a solution are the **solvent** and the **solute.** The dissolved substance is called a **solute** and the substance, which dissolves the solute, a **solvent**. The solvent is present in a larger proportion than the solute. Classification of solutions can be based on various criteria as follows:

1. Degree of Saturation

Saturated solution: A solution in which no more solute can be dissolved at a particular temperature is called a saturated solution.

Unsaturated solution: A solution in which more quantity of solute can be dissolved at a given temperature is called unsatured solution.

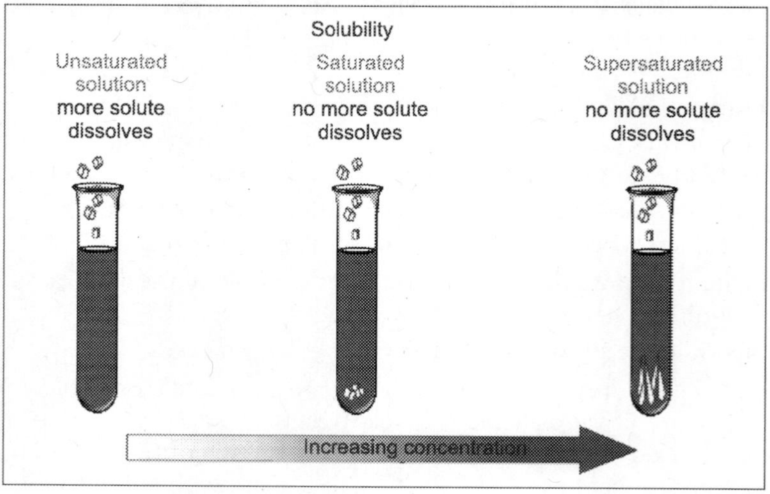

2. Particle Size of the Solute

True Solution: A homogenous solution in which the particle size ranges between 10^{-10} to 10^{-9} m (or it is less than 10^{-7} cm) e.g., a solution of common salt in water.

Colloidal solution: A colloidal solution is a "two-phase heterogeneous system in which a substance is distributed in colloidal state (particles having diameter between 1 to 100 nm) in a medium". The particles of the dispersed substance (of colloidal size) are called dispersed phase; while the medium on which they are dispersed, is called dispersion medium, e.g., milk, butter, smoke, etc.

Suspension: It is a heterogeneous mixture which contains small insoluble particles of the solute spread throughout the solvent without dissolving in it. The particle size of the solute is greater than 10^{-5} cm, e.g., muddy water, paint, etc.

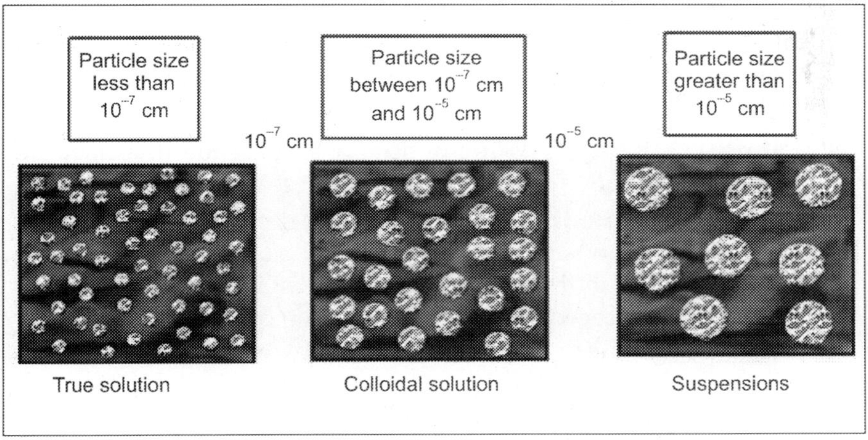

3. Stability of the Standard Solution

A solution, whose concentration is known, is called a **standard solution**. There are two types of standard solutions.

1. *Primary standard solution*

A solution that is prepared directly by dissolving a definite weight of a substance, in a definite volume is called a **primary standard**. The properties of primary standard are following:

(a) It is easily available in state of high purity.
(b) It is neither hygroscopic nor deliquescent.
(c) It shows high solubility in water.
(d) It does not dissociate or decompose during storage.

2. *Secondary standard solution*

A solution that cannot be prepared directly by weighing and its exact strength can only be determined by titrating it with the primary standard solution (i.e., by standardization) is called **secondary standard**.

TITRATION

The process of finding out the concentration of a substance in solution by reacting it with another solution of known concentration is called **volumetric analysis.** It involves addition of one solution from burette to another in the conical flask, in order to complete the chemical reaction between the two solutions (i.e., **titration).** Following are the various types of titrations commonly used in food analysis:

1. **Acid Base Titration:** In this type, an acid is titrated with a base of known strength or vice versa to find the strength of the other.
2. **Redox Titration:** In this type, a reducing agent is titrated against oxidizing agent or vice versa and such titration involving redox reactions are termed redox titrations, e.g., of oxidizing agents: $KMNO_4$, $K_2Cr_2O_7$., e.g., of reducing agents: oxalic acid, Mohr's salt.
3. **Iodine Titrations:** The redox titrations using iodine directly or indirectly are called iodine titrations. These are of two types.
   (a) **Iodimetric:** Iodimetric titrations are defined as those iodine titrations in which a standard iodine solution is used as an oxidant and iodine is directly titrated with a reducing agent. Iodimetric procedures are used for determination of reducing agents like thiosulfates, sulfites, etc., by titrating them against standard solution of iodine run in from a burette.
   (b) **Iodometric:** These are defined as those iodine titrations in which some oxidizing agent liberates iodine from iodine and then liberated iodine is titrated against standard solution of a reducing agent added from a burette. In such titrations, a neutral or an acidic solution of oxidizing agent is employed. The amount of iodine liberated from iodine (KI) is equivalent in quantity of oxidizing agent present.

End Point: The stage during titration at which the reaction is just complete is known as the end point of the titration.

Indicator: The reagent used to locate the exact completion stage, i.e., end point of a reaction by showing change in its color is called 'indicator'.

External Indicators: It is placed outside on a reference porcelain plate, e.g., $K_3Fe(CN_6)$ solution in titration of ferrous sulfate or ferrous ammonium sulfate against $K_2Cr_2O_7$.

Self Indicator: When color of one of the titrants acts as indicator, e.g., pink color of permanganate ion in titration of oxalic acid.

Internal Indicator: It is added to solution taken in the conical flask or beaker, e.g., phenolphthalein and methyl orange.

A good indicator must possess the following two essential characteristics:

(i) The color change of the indicator must be clear and sharp, i.e., must be sensitive.

(ii) The pH range over which the color changes take place must be such as to indicate when the reaction is complete.

MATERIALS REQUIRED

Apparatus: Weighing balance

Glassware: Beakers, Conical Flask, Burette, Funnel, Measuring Cylinder

Reagents:

1. Hydrochloric acid
2. Sodium thiosulfate
3. Sodium hydroxide
4. Oxalic acid
5. Potassium dichromate
6. Phenolphthalein
7. Starch

PROCEDURE

Preparation of Standard Solutions

1. 0.1 N Sodium Thiosulfate

Take 25 gm (24.82 gm) of $Na_2S_2O_3.5H_2O$ in a volumetric flask and make up the volume up to 1 L with distilled water. Mix thoroughly and allow it to stand overnight. Siphon off the clear liquid. Standardize it with $K_2Cr_2O_7$.

To 0.20 to 0.23 gm of $K_2Cr_2O_7$ (dried for 2 hours at 105°C), add 150 ml distilled water, 2 gm KI and mix. Now add 20 ml of 1 N HCl. Swirl and allow it to stand for 10 minutes. Titrate it with $Na_2S_2O_3$ from burette adding about 80% of the solution first and then adding using 1 ml of 1% starch solution as an indicator. The end point is appearance of light color.

$$\text{Normality} = \frac{\text{Weight of } K_2Cr_2O_7 \times 1000}{\text{ml of } Na_2S_2O_3 \times 49.037}$$

2. 0.1 N HCl

Molecular weight of HCl is 36.46 gm. Hence, 0.1 N contains 3.646 gm HCl/L.

Rest of the calculations are as per the purity (p) and specific gravity (g) mentioned on the reagent bottle so for preparation of 1000 ml of 0.1 N HCl the formula is

$$\frac{3.646 \times 100}{p \times g}$$

3. 0.1 N NaOH

Molecular weight of NaOH is 40 gm. Hence, 0.1 N NaOH can be prepared by dissolving 4 gm of NaOH in 1 liter of distilled water and standardizing using 0.1 N oxalic acid with phenolphthalein as an indicator.

4. 0.1 N Oxalic Acid

Weigh 6.3023 gm of oxalic acid and make up the volume to 1L with distilled water.

CALCULATIONS

I. Preparation of Solutions

(i) 0.1 N HCl

1000 ml of 1 N HCl contains 36.46 gm of HCl

1000 ml of 0.1 N HCl contains = 36.46 × 0.1

= 3.646 gm

$$\text{Volume of HCl required} = \frac{\text{wt.} \times 100}{\text{Sp gravity} \times \% \text{ purity}}$$

(ii) 0.1 N $Na_2S_2O_3$

$$\text{Amount of } Na_2S_2O_3 \text{ required} = \frac{N \times \text{Eq.wt} \times 1000}{1000}$$

$$= \frac{0.1 \times 248.7 \times 1000}{1000}$$

= 24.87 g/L

(iii) 0.1 N Oxalic Acid

Equivalent weight of oxalic acid

$$= \frac{\text{Molecular weight}}{\text{No. of electrons lost by one molecule}}$$

$$= \frac{126}{2} = 63$$

1 N oxalic acid solution contains 63 gm of oxalic acid/1000 ml water and 0.1 N oxalic acid solution = 6.3 gm of oxalic acid/1000 ml water

II. Standardization of solutions

For example, 0.1 N HCl

Titration of HCl v/s NaOH

S.No	*Initial burette reading (ml)*	*Final burette reading (ml)*	*Volume used (final – initial) (ml)*

Volume of NaOH used =

Normality of NaOH (N_1) =

Volume of NaOH used (V_1) =

Normality of HCl (N_2) =

Volume of HCl used (V_2) =

$$N_1V_1 = N_2V_2$$

$$N_2 = \frac{N_1V_1}{V_2}$$

Note: Follow similar procedure for sodium thiosulfate with potassium dicromate

OBSERVATIONS

S. No.	*Solution*	*Normality (N)*	
		Theoretical value	*Experimental value*

RESULT

The normality of the solution prepared should be close to the desired normality. The discrepancy in result if any, should be analyzed in terms of water quality and the purity of the solute added.

PRECAUTIONS

1. All the glassware should be properly cleaned.
2. Always add acid in water, not water in acid.
3. End point should be accurately noted.
4. Weighing should be done carefully.
5. Three concordant titer values should be taken.

REVIEW QUESTIONS

1. Name some primary standard solutions.
2. Why $Na_2S_2O_3$ is a secondary solution?
3. How will you prepare 20 ppm of sodium hydroxide solution?
4. Differentiate between direct and back titrations giving examples.
5. State different types of solutions based on their particle size.

5
Component Analysis

Food analysis is the discipline that deals with the development, application and study of analytical procedures for characterizing the properties of foods and their constituents. These analytical procedures are used to provide information about a wide variety of different characteristics of foods, including their composition, structure, physicochemical properties and sensory attributes. This information is critical to our rational understanding of the factors that determine the properties of foods, as well as to our ability to economically produce foods that are consistently safe, nutritious and desirable and for consumers to make informed choices about their diet. The objective of this manual is to review the basic principles of the analytical procedures commonly used to analyse foods and to discuss their applications to specific food components, e.g., lipids, proteins, water, carbohydrates and minerals.

The composition of a food largely determines its safety, nutrition, physicochemical properties, quality attributes and sensory characteristics. Most foods are compositionally complex materials made up of a wide variety of different chemical constituents. Their composition can be specified in a number of different ways depending on the property that is of interest to the analyst and the type of analytical procedure used: specific atoms (e.g., carbon, hydrogen, oxygen, nitrogen, sulfur, sodium, etc.); specific molecules (e.g., water, sucrose, tristearin, β-lactoglobulin), types of molecules (e.g., fats, proteins, carbohydrates, fiber, minerals), or specific substances (e.g., peas, flour, milk, peanuts, butter). Government regulations state that the concentration of certain food components must be stipulated on the nutritional label of most food products, and are usually reported as specific molecules (e.g., vitamin A) or types of molecules (e.g., proteins).

The structural organization of the components within a food also plays a large role in determining the physicochemical properties, quality attributes and sensory characteristics of many foods. Hence, two foods that have the same composition can have very different quality attributes if their constituents are organized differently. There are usually a number of different analytical techniques available to determine a particular property of a food material. It is therefore, necessary to select the most appropriate technique for the specific application. The analytical technique selected depends on the property to be measured, the type of food to be analyzed, and the reason for carrying out the analysis. Information about the various analytical procedures available can be obtained from a number of different sources. An analytical procedure may already be routinely used in the laboratory or company where you are working.

Food products are analyzed for a variety of reasons, e.g., compliance with legal and labeling requirements, assessment of product quality, determination of nutritive value, detection of adulteration, research and development. The sections will cover the basic principles of analytical procedures and techniques commonly used to provide information about the chemical composition, structure and physical properties of food materials. The aim of the laboratory manual is to give students experience in performing food analysis experiments, analyzing data and reporting their findings.

5.1 MOISTURE

Moisture content is one of the most commonly measured properties of food materials. It is important to food scientists for a number of different reasons:

Legal and Labeling Requirements: There are legal limits to the maximum or minimum amount of water that must be present in certain types of food.

Economic: The cost of many foods depends on the amount of water they contain – water is an inexpensive ingredient, and manufacturers often try to incorporate it as much as possible in a food, without exceeding some maximum legal requirement.

Microbial Stability: The propensity of microorganisms to grow in foods depends on their water content. For this reason, many foods are dried below some critical moisture content.

Food Quality: The texture, taste, appearance and stability of foods depend on the amount of water they contain.

Food Processing Operations: Knowledge of the moisture content is often necessary to predict the behavior of foods during processing, e.g., mixing, drying, flow through a pipe or packaging.

It is therefore, important for food scientists to be able to reliably measure moisture contents. A number of analytical techniques have been developed for this purpose, which vary in their accuracy, cost, speed, sensitivity, specificity, ease of operation, etc. The choice of an analytical procedure for a particular application depends on the nature of the food being analyzed and the reason that information is needed.

The moisture content of a food material is defined through the following equation:

$$\% \text{ Moisture} = (m_w / m_{\text{sample}}) \times 100$$

where m_w is the mass of the water and m_{sample} is the mass of the sample.

Despite having the same chemical formula (H_2O), the water molecules in a food may be present in a variety of different molecular environments depending on their interaction with the surrounding molecules.

Bulk water: Bulk water is free from any other constituents, so that each water molecule is surrounded only by other water molecules.

Capillary or trapped water: Capillary water is held in narrow channels between certain food components because of capillary forces. Trapped water is held within spaces in a food that are surrounded by a physical barrier which prevents the water molecules from easily escaping, e.g., an emulsion droplet or a biological cell.

Physically bound water: A significant fraction of the water molecules in many foods are not completely surrounded by other water molecules, but are in molecular contact with other food constituents, e.g., proteins, carbohydrates or minerals.

Chemically bound water: Some of the water molecules present in a food may be chemically bonded to other molecules as water of crystallization or as hydrates, e.g., $NaSO_4.10H_2O$.

The thermal energy used to evaporate the water from a food sample can be provided directly (e.g., transfer of heat from an oven to a food) or indirectly (e.g., conversion of electromagnetic radiation incident upon a food into heat due to absorption of energy by the water molecules). The most commonly used methods are:

OVEN DRYING METHODS

With oven drying, the sample is heated under specified conditions, and the loss of weight is used to calculate the moisture content of the sample.

Forced Oven Draft: Sample is rapidly weighed into a moisture pan and placed in the oven for an arbitrarily selected time if no standard method exists. Drying time periods for this method are 0.75-24 hours, depending on the food sample.

Vacuum Oven: Drying under reduced pressure (25-100 mm Hg) allows a more complete removal of water and volatiles without decomposition within 3-6 hrs drying time.

Microwave Oven: A precise and rapid technique that allows some segments of the food industry to make in-process adjustments of moisture content before final packaging. In vacuum microwaves, a drying time of 10 minutes can yield results equivalent to those of five hours in a standard vacuum oven.

Infrared Drying: Employs penetration of heat into the sample being dried, as compared to heat conductivity and convection as with conventional ovens. The required drying time can be as little as 10-25 minutes (suited for qualitative in-process use, but not approved by the AOAC).

DISTILLATION METHODS

Distillation methods are based on *direct* measurement of the amount of water removed from a food sample by evaporation: % Moisture = 100 ($M_{WATER}/M_{INITIAL}$). In contrast, evaporation methods are based on *indirect* measurement of the amount of water removed from a food sample by evaporation: %Moisture = 100 ($M_{INITIAL}$ - M_{DRIED})/$M_{INITIAL}$. Basically, distillation methods involve heating a weighed food sample ($M_{INITIAL}$) in the presence of an organic solvent that is immiscible with water. The water in the sample evaporates and is collected in a graduated glass tube where its mass is determined (M_{WATER}).

CHEMICAL METHODS

Karl Fischer Titration

This technique is particularly suited to food products that show erratic results when heated or submitted to vacuum. It is the method of choice for low-moisture foods such as dried fruits and vegetables, candies, chocolate, roasted coffee, oils and fats, and low-moisture foods high in sugar or protein.

PHYSICAL METHODS

Electric (dielectric or conductivity): Moisture content is determined by measuring the change in capacitance or resistance to an electric current passed through a sample.

Hydrometry: Used to determine moisture/ solid content of beverages and sugar solutions. Measuring the specific gravity or density of the sample via one of the following instruments:

(i) *Pycnometer:* used to compare the weights of equal volumes of a liquid and water. It yields density of the liquid compared to water.

(ii) *Hydrometer:* a standard weight on the end of a spindle which displaces a weight of liquid equal to its own weight. In a low-density liquid, the weight will sink to a greater depth.

(*iii*) *Westphal Balance:* functions on the principle that the plummet on the balance will be buoyed by the weight of liquid equal to the volume displaced.

Refractometry: Measures moisture content of oils and syrups as a function of the degree of refraction of a light beam as it passes through the sample.

Infrared Analysis: Measures the energy that is reflected or transmitted by the sample when exposed to infrared light.

Freezing Point: Measures the solutes present by determining the freezing point of the sample. Used principally to measure for added water content

DIFFERENT TYPES OF METHODS FOR MOISTURE ESTIMATION AT A GLANCE

1. Evaporation Methods
    - Convection and force draft ovens
    - Vacuum oven
    - Microwave oven
    - Infrared lamp drying
2. Distillation Method
    - Dean and Stark Method
3. Chemical Reaction Methods
    - Karl Fischer Method
    - Gas Production Method
4. Physical Method
    - Density Method
5. Spectroscopic Method
    - Nuclear Magnetic Resonance (NMR)
6. Vapor Pressure Methods
7. Thermogravimetric Methods
8. Calorimetric Methods

5.1.1 Determination of Moisture Content using Hot Air Oven Method

INTRODUCTION

The method is used for the quantitative determination of moisture in all foods except food samples high in sugar and fat (> 10%). Moisture in this method refers to the amount of free water and volatile substances that are lost by drying the food under controlled temperature in an air oven. It is expressed in g per 100 g sample. The method is based on the drying of food sample under controlled temperature until constant weight is obtained. Moisture content is required to express the nutrient content per dry weight basis. In some foods, moisture is used to indicate their quality. Standard values of moisture are indicated in food notification or regulation.

Table showing Moisture content of selected foods

Food item	*% moisture (wet wt. basis)*
CEREALS AND PASTA	
Wheat flour, whole grain	10.3
Corn flakes cereal	3.0
Macaroni, dry, enriched	10.2
DAIRY PRODUCTS	
Yogurt, plain, low fat	89.0
Cheddar cheese	37.5
Ice cream, vanilla	61.0
FATS AND OILS	
Butter, salted	16.9
Oil, soybean	0
FRUITS AND VEGETABLES	
Watermelon, raw	91.5
Apples, raw, with skin	83.9
Raisins	15.4
Cucumbers, with peel, raw	96.0
Potatoes, raw, flesh and skin	79.0
MEAT, POULTRY, AND FISH	
Beef, ground, extra lean, raw	63.2
Chicken, light meat, meat and skin, raw	68.6
Finfish, flounder/sole, raw	79.1
Egg, whole, raw, fresh	75.3

Weighed samples are placed in an oven for a specified time and temperature (e.g., 3 hours at 100°C) and their dried mass are determined, or they are dried until they reach constant mass. There are often considerable temperature variations within convection ovens, and so precise measurements are carried out using forced draft ovens that circulate the air so as to achieve a more uniform temperature distribution within the oven.

Table showing drying conditions for moisture determination of various food products

Product	*Predry on steam bath*	*Temperature (°C)*	*Pressure (mm Hg)*	*Drying Time (h)*
Cheese, high moisture (soft, processed)	Yes	100	<100	4
Cheese, low moisture		100	<100	4
Coffee, roasted		98-100	<25	5.5
Coffee, roasted		70	<37.5	16
Egg, dried		98 - 100	<25	5
Egg, liquid	Yes	98 - 100	<25	5
Flour		98 - 100	<25	5
Flour		130	Atmospheric	1–2
Fruit, dried		70	<100	6
Ice cream	Yes	100	Atmospheric	3.5
Meat		100 - 102	Atmospheric	16–18
Milk, fluid	Yes	98 - 100	Atmospheric	3
Milk, dried		100	<100	5
Nuts		95 - 100	<100	5

MATERIALS REQUIRED

Apparatus: Moisture dishes, Drying oven, Weighing balance, Desiccators, Spatula, Hot air oven

Sample/s Used: Food Samples— Bread, Wheat grains, Potato, Oil

PROCEDURE

1. Weigh 5 g of sample in a clean and tarred moisture dish in duplicates (in some cases 1 g or 10 g of the sample can be taken).
2. Spread the sample uniformly in the dish and place the dish in the hot air oven maintained at 100°C with the open lid for about 5 hrs.

3. The dish is then removed from oven and lid closed and cooled in a desiccator.
4. Cooled dish is then weighed (repeat for a constant weight).
5. Calculate the moisture content using the formula given below.

FORMULA USED

$$\text{*Moisture content (\%)} = \frac{W_2 - W_3}{W_2 - W_1} \times 100$$

where

W_1 = Weight of the moisture dish (g)

W_2 = Weight of moisture dish + sample (before drying) (g)

W_3 = Weight of moisture dish + sample (after drying) (g)

$$\text{Moisture content (\%)} = \frac{\text{Weight of moisture in the sample}}{\text{Weight of wet sample}} \times 100$$

Total solid (%) = 100 – % moisture (w/w)

Note: Report test results (in g per 100 g sample) to one decimal place.

OBSERVATIONS

The moisture content of the given samples is tabulated below:

S. No.	*Sample*	*W_1(g)*	*W_2 (g)*	*W_3 (g)*	*%M.C.**	*AVG.%M.C.*

* Formula used

Table showing experimental values and theoretical values of moisture of different samples:

S. No.	*Sample*	*Experimental moisture content*	*Standard moisture content (%)*	*Reference*

RESULTS

The rate and extent of moisture removal depends on the size and shape of the sample, and how finely it is grounded. The greater the surface area of material exposed to the environment, the faster the rate of moisture removal.

Some samples tend to clump together or form a semi-permeable surface crust during the drying procedure. This can lead to erroneous and irreproducible results because the loss of moisture is restricted by the clumps or crust. For this reason samples are often mixed with dried sand to prevent clumping and surface crust formation. In fact, all the experimental values of moisture obtained should be compared with the literature values to study any deviations.

PRECAUTIONS

1. The moisture dishes should be tarred to a constant weight before use.
2. The samples should not be exposed to air for a long time.

5.1.2 Determination of Moisture Content using Karl Fischer Method

INTRODUCTION

Another direct method for moisture analysis is based on the reduction of iodine by sulfur dioxide in the presence of water, as proposed by Bunsen in 1853:

$$2H_2O + SO_2 + I_2 \longrightarrow H_2SO_4 + 2HI$$

Fischer modified the conditions of the reaction, allowing quantification of moisture. This method has been then known as the Karl Fischer titration method. The Karl Fischer method has become a standard method for the moisture measurement in liquids and solids due to its selectivity, high precision, and speed. The method is particularly applicable for determining moisture in foods for which heating methods give erratic results. Moisture determination using this chemical technique has been approved for dried vegetables, oils and fats, cocoa products, and liquid molasses. The sensitivity of this method is superior to that of other methods, being able to quantify the amount of water to a few parts per million.

The reagent for the Karl Fischer titration method consists of iodine, pyridine, sulfur dioxide, and methanol. The titration of water with this reagent follows the two-step reaction shown below.

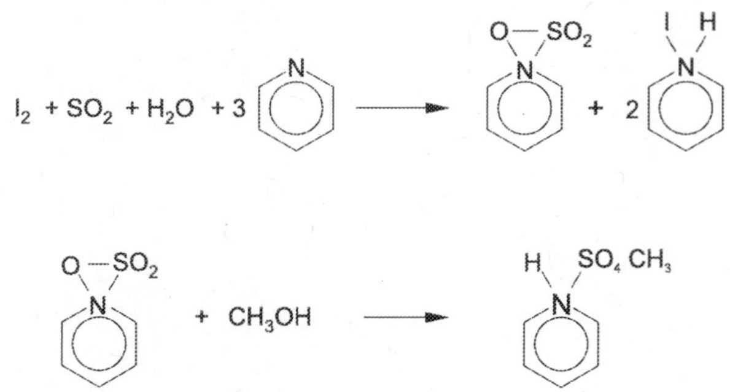

As indicated by the above reactions, titration of 1 mol of water requires 1 mol of iodine, 1 mol of sulfur dioxide, 3 mol of pyridine, and 1 mol of methanol. The titration is conducted either by volumetric titration, where the end point is indicated by the appearance of brown color from free iodine (detected either visually or by photometric determination), or by coulometric titration, where the end point is determined by a potentiometer.

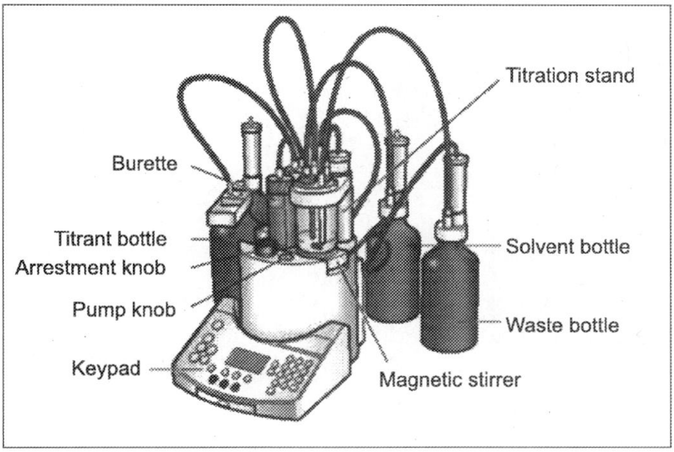

There are **two methods** used to perform the **Karl Fischer titration test.**

One is known as **Volumetric Karl Fischer.** With this method the moisture determination is based on the amount, or volume, of reagent used to convert the water. In the **volumetric method,** samples are dissolved in a solvent before the titration begins. A reagent is added until the water is removed.

The other method is known as **Coulometric Karl Fischer**. In this approach, the reagent and solvent are combined in the titration cell. *When a sample is introduced into the titration cell and dissolved, reagent is released by the induction of an electrical current.* The amount of current required to convert the water is the determinant of the amount of moisture. A **Coulometric Karl Fischer** instrument is often referred to as a **coulometer.**

The advantage of the Coulometric Karl Fischer method is the capability to accurately measure small amounts of moisture. Sensitivity of these instruments is as low as 0.1 microgram (µg) of water. **This method is normally used for moisture content below 1** % or for samples where the moisture is less than 200 micrograms.

MATERIALS REQUIRED

Apparatus:

1. Burette: automatic filling type, all glass, fully protected against moisture ingress.
2. Titration vessel. Having an agitation device, such as magnetic stirrer, slightly pressurized with dry inert gas (N_2 or CO_2) to exclude air.
3. Electrometric apparatus and galvanometer, suitable for "dead stop" end point technique.

Reagents:

1. **Methanol:** anhydrous.
2. Karl Fischer reagent consists of a solution of iodine in methanol, and a solution of sulfur dioxide in pyridine. Both the solutions are mixed shortly before use.

Sample/s Used: Skim milk powder, Gram flour

PROCEDURE

1. Quickly weigh an amount of sample containing approximately 100 mg water into a predried round bottomed 50 ml flask.
2. Add 40 ml methanol into the flask, quickly place the flask on the heating source, and connect the reflux condenser.
3. Boil the contents of the flask gently under reflux for 15 min.
4. Stop heating with the condenser attached, and let it drain for 15 min.
5. Remove and stopper the flask.
6. Pipette a 10 ml aliquot of the extract into the titration vessel, titrate with the Karl Fischer reagent to the "dead stop" end point and record volume of titrant used.
7. Run a blank flask without sample, following the same procedure described above.

CALCULATIONS

For this protocol, moisture content is be calculated using the following equation. If a known sample mass was introduced directly into the titration vessel rather than extracting with methanol, the dilution factor (i.e., 4) would be eliminated from the equation.

$$\% \text{ moisture} = \frac{1}{W}[(4 \times 0.1 \times F)\ \text{X}\ (A - B)]$$

where

W = Weight of the sample

A = ml reagent used for sample

B = ml reagent used for blank

F = standardization factor of the reagent

From titrating samples containing known amounts of water, the standardization factor of the reagent, F, can be determined, which has the units of mg water/ml reagent.

PRECAUTIONS

The reagents and solutions used in the determination of water by this method are sensitive to water and precautions must be taken throughout to prevent exposure to atmospheric moisture.

5.1.3 Determination of Moisture Content using Dean and Stark Method

INTRODUCTION

Distillation techniques involve co-distilling the moisture in a food sample with a high boiling point solvent that is immiscible in water, collecting the mixture that distills off, and collecting the distillate in a graduated tube. One of the most known distillation methods is the Dean and Stark method. A known weight of food is placed in a flask with an organic solvent such as xylene or toluene. The organic solvent must be insoluble with water, have a higher boiling point than water, be less dense than water, and be safe to use. The flask containing the sample and the organic solvent is attached to a condenser by a side arm and the mixture is heated. The water in the sample evaporates and moves up into the condenser where it is cooled and converted back into liquid water, which then trickles into the graduated tube. When no more water is collected in the graduated tube, distillation is stopped and the volume of water is read from the tube.

MATERIALS REQUIRED

Apparatus: Moisture distillation apparatus consists of a 500 ml short neck round bottomed flask heated by suitable means and provided with a reflux condenser discharging into a trap connected to the flask. The trap serves to collect and measure the condensed water and to return the condensed solvent to the flask (refer Fig. 2) and a copper wire long enough to extend throughout the condenser with one end twisted into a spiral. The diameter of the spiral should be such that it fits snugly within the graduated portion of the receiver and yet, it can be moved up and down.

Reagents: Toluene – saturated with small quantity of water and distilled. Use the distillate for determination of moisture.

Sample/s Used: Spices such as cardamom, cinnamon

PROCEDURE

1. Clean the entire apparatus with chromic acid cleaning solution to minimize adherence of water droplets to the sides of the condenser and the receiver.
2. Rinse thoroughly with water and dry completely before use.
3. Weigh to the nearest 0.01 gm about 20-40 gm of prepared sample (or enough to yield 2-5 ml H_2O in the trap).
4. Transfer to the distilling flask with toluene. Add enough toluene to cover test portion completely (about 75 ml). Fill receiving tube

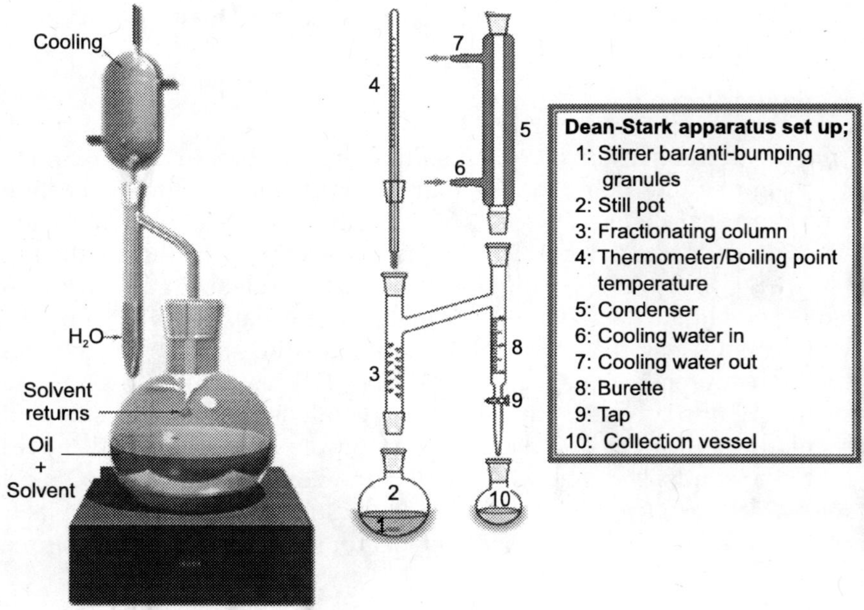

Dean-Stark method set up

with toluene, pouring it through top of the condenser until it begins to overflow into the distillation flask.

5. Add a few glass beads to avoid bumping. Bring to boil and distill slowly about 2 drops per second (about 100 drops per minute) until most of the water distills over, then increase the rate of distillation to 4 drops per second. Continue distilling until 2 consecutive readings 15 minutes apart show no change.
6. Rinse condenser carefully with 5 ml toluene. Continue distillation for 3-5 minutes. Then, cool the receiver to room temperature allowing it to stand in air or cooling it in water. Solvent and water layers should now be clear, if not, let it stand until clearing occurs.

OBSERVATIONS

Read volume of water estimating to nearest 0.1 ml and calculate percentage.

$$\text{Moisture content (\% by weight)} = \frac{100\,V}{M}$$

where V = Volume in ml of water collected

M = Weight of sample

RESULTS

The moisture content (% by weight) of the given sample is found to be_________.

PRECAUTIONS

1. Water droplets should not adhere to the inside of the glassware.
2. Avoid decomposition of thermally labile samples at the elevated temperatures as it leads to erroneous results.
3. Insert loose cotton plug in top of the condenser to prevent condensation of atmospheric moisture in the tube.
4. Dislodge any water held up in the condenser with the wire loop.

5.1.4 Determination of Moisture Content using Vacuum Oven Drying Technique

INTRODUCTION

Vacuum-oven drying is generally the standard and most accurate drying method for moisture analysis of foods. Many of the drawbacks associated with air-oven drying can be overcome by this method. Typically, vacuum drying heats foods at 98–102°C and 25–100 mm Hg pressure. Lower temperatures (60–70°C) can be used for products high in sugar to prevent decomposition. At the reduced pressure, moisture evaporates quickly and drying times can be dramatically reduced.

MATERIALS REQUIRED

Apparatus:

1. **Vacuum oven:** Thermostatically controlled and connected to a vacuum pump capable of maintaining the pressure in the oven below 25 mm Hg. The oven should have a dry-air inlet that passes through an indicating desiccant and trap for releasing the vacuum.
2. **Moisture Dishes:** Aluminium dishes with close-fitting lids and flat bottoms to provide maximum area of contact with the heating plate.
3. Other apparatus and equipment are the same as those for air-oven drying.

Sample/s Used: Butter, vegetable oil

PROCEDURE

1. Wash and dry the aluminum sample dishes. After cooling in a desiccator, weigh the dish to 0.1 mg.
2. Weigh a 3.0–5.0 g sample into the preweighed dish using an analytical balance. Distribute the sample evenly over the bottom of the dish. Some samples require predrying as described for air-oven drying to prevent decomposition and splattering.
3. Place the sample dishes in the vacuum oven, partially uncover the dish, evacuate the oven, and dry the sample at an appropriate temperature and vacuum pressure. During drying, admit a slow current of air dried by passage through indicating desiccant into the oven.
4. Shut off the vacuum pump after 5 h and slowly release the dry air into the oven. Press cover lightly on to dish using tongs, transfer the dish to a desiccator to cool, and reweigh.

5. Dry for another hour to ensure that constant weight has been achieved.
6. Calculations same that of the air oven drying method.

RESULTS

The moisture content of the given sample is found to be________.

PRECAUTIONS

1. Spread the sample uniformly over the surface of the moisture dish.
2. Do not cover the moisture dish with the lid inside the vacuum oven.
3. Keep the moisture dish inside the desiccator for cooling and then weigh.

REVIEW QUESTIONS

1. What is Dean and Stark method?
2. Give the principle of air oven method.
3. What solvent is used in Dean and Stark method?
4. Why is Dean and Stark method useful for volatile constituents?
5. What method is employed for moisture content estimation in cereals and pulses?
6. What is the time temperature combinations used for low and high moisture foods for moisture estimation?
7. What is the principle of Karl Fischer Method?
8. What is the advantage of vacuum drying method over air oven method?
9. Why is the milk sample partially evaporated on a hot plate before being dried in the hot air oven?
10. What is the difference between moisture content and water activity measurements?

5.2 ASH AND MINERALS

Although the amounts of mineral elements present in food are generally small (often less than 1% of the food in total), the amounts of individual elements will vary considerably depending on the particular element and on the origin of the food, e.g., the soil in which the plant food material has been grown or the composition of the diet of the animal from which the food is produced. Based on the typical levels of mineral elements found in foods, they may be classified on the following basis:

MAJOR ELEMENTS (> 0.01 % or 100 ppm)			
Calcium	Chlorine	Magnesium	Sulfur
Phosphorus	Potassium	Sodium	
TRACE ELEMENTS (< 0.01% or 100 ppm)			
Arsenic	Chromium	Cobalt	
Copper	Fluorine	Iodine	
Iron	Manganese	Molybdenum	
Nickel	Selenium	Silicon	
Tin	Vanadium	Zin	

In addition to the above elements, which are all essential in the human diet, certain other elements may be found in foods, usually in trace amounts, which in addition to being non-essential are also toxic. These include:

Beryllium	Cadmium	Mercury
Lead	Palladium	Thallium

The ash content of a foodstuff is the residue remaining after all the moisture has been removed and the organic material (fats, proteins, carbohydrates, vitamins, organic acids, etc.) has been burnt away by igniting at a temperature of around 500°C. This results in the oxidation of organic constituents to volatile materials such as carbon dioxide, nitrogen oxide and sulfur dioxide. The residue is generally considered to be a measure of the mineral content of the original food although differences in composition between the ash content and the mineral matter may occur, as a result of the loss of some volatile inorganic constituents such as chlorides during the ashing process and the presence in the ash residue of constituents originating from organic food constituents, such as sulfur from protein.

The general procedure for estimating the ash content is to weigh a known amount of the food into a previously ignited and weighed silica dish, to char the food using a low flame or infrared lamp, and then to ignite overnight in a muffle furnace at a temperature of 500-550°C. Where difficulties occur with the ashing process and incomplete oxidation occurs, resulting in a black residue indicative of incompletely oxidised carbon, rather than a desirable grey ash, the residue may be moistened with acid, dried and re-ignited.

Individual mineral elements may be estimated by taking the ash, boiling it with concentrated hydrochloric acid to solubilise the mineral elements, diluting to known volume with water, and then estimating individual minerals by methods such as titrations for chloride, colorimetry for phosphorus and atomic absorption spectrophotometry for most metals.

Thus, the *"ash content"* is a measure of the total amount of minerals present within a food, whereas the *"mineral content"* is a measure of the amount of specific inorganic components present within a food, such as Ca, Na, K and Cl. Determination of the ash and mineral content of foods is important for a number of reasons such as:

Nutritional labeling: The concentration and type of minerals present must often be mentioned on the label of a food.

Quality: The quality of many foods depends on the concentration and type of minerals they contain, including their taste, appearance, texture and stability.

Microbiological stability: High mineral contents are sometimes used to retard the growth of certain microorganisms.

Nutrition: Some minerals are essential to a healthy diet (e.g., calcium, phosphorus, potassium and sodium) whereas others can be toxic (e.g., lead, mercury, cadmium and aluminum).

Processing: It is often important to know the mineral content of foods during processing because this affects the physicochemical properties of foods.

5.2.1 Determination of the Total Ash Content, Acid Soluble and Insoluble Ash and Water Soluble and Insoluble Ash

INTRODUCTION

Ash content refers to the total mineral residue left after incineration of organic matter in a food stuff. It has no nutritional significance per se, but the value for ash is a useful check, in summing up the proximate composition of food and a measure of its mineral content. The method involves oxidation of all organic matter by incineration in a muffle furnace at a specified temperature (<550°C). Ashing above 650°C will volatilise inorganic salts like alkali chloride and a portion of ash will fuse and enclose some carbon, preventing them from being ignited. The residue left after incineration is the **ash content** of the sample.

A basic knowledge of characteristics of various ashing procedures and types of equipment is essential to ensure reliable results. The ash content of foods can be expressed on either a wet weight or on a dry weight basis. Three major types of ashing are available:

- Dry ashing for majority of samples.
- Wet ashing (oxidation) for samples with high fat content (meat and meat products) and as a preparation for elemental analysis.
- Low temperature plasma dry ashing for preparation of samples when volatile elemental analysis is conducted.

Table showing different types of ashing techniques

Dry ashing	**Advantages**	**Disadvantages**
Refers to the use of a muffle furnace capable of maintaining temperature of 500–600°C. Water and volatiles are vaporized and organic substances are burned in presence of O_2 in air to CO_2 and oxides of N_2. Most minerals are converted to oxides, sulfates, phosphates, chlorides and silicates. Elements such as Fe, Se, Pb and Hg may partially volatilize.	Safe, few reagents are required, many samples can be analyzed simultaneously, not labor intensive and ash can be analyzed for specific mineral content.	Long time required (12-24 hours), muffle furnaces are quite costly to run due to electrical costs, loss of volatile minerals at high temperatures, e.g., Cu, Fe, Pb, Hg, Ni, Zn.
Wet ashing	**Advantages**	**Disadvantages**
Refers to a procedure for oxidizing organic substances by using acids and oxidizing agents or their combinations. Minerals are solubilized without volatilization. Wet ashing is often preferable to dry ashing as a preparation for specific elemental	Little loss of volatile minerals occurs because of lower temperatures used, more rapid than dry ashing.	Labor intensive, requires a special fume-cupboard if perchloric acid is used because of its hazardous nature, low sample output.

analysis. Nitric and perchloric acids are preferable, but a special perchloric acid hood is essential. Caution must be taken when fatty foods are used.		
Low temperature plasma ashing Refers to a specific type of dry ashing method whereby foods are oxidized in a partial vacuum by nascent oxygen formed by an electromagnetic field. Ashing occurs at a much lower temperature than in a muffle furnace, preventing volatilization of most elements. The crystalline structure usually remains intact.	**Advantages** Less chance of losing trace elements by volatilization	**Disadvantages** Relatively an expensive equipment and small sample output.

MATERIALS REQUIRED

Apparatus: Muffle furnace,Weighing balance, Silica crucibles and Ashless filter paper (Whatman No. 40).

Reagent: 10% HCl.

Sample/s Used: Food Samples: Wheat flour, Soyabean.

PROCEDURE

I. Determination of Total Ash

1. Weigh 5-10 g of sample in a tarred crucible. Predry if the sample is very moist.
2. Ignite the sample on flame to remove organic materials.
3. Place the crucible in muffle furnace.
4. Ignite for 12-18 hours at about 550°C.
5. Turn off the muffle furnace and take out the crucible when temperature drops to 5°C or lower.
6. Transfer the crucible to a dessicator for cooling. Then weigh and calculate the total ash.

II. Determination of Water Insoluble Ash

1. Add 25 ml distilled water to the dry ash obtained.
2. Cover the crucible with watch glass and heat nearly to boiling.
3. Filter through an ashless filter paper and rinse with hot distilled water several times.
4. Dry and reash the filter paper at 500°C.
5. Weigh and calculate as % water insoluble ash. Then substract this from total ash to obtain % water soluble ash.

III. Determination of Acid Insoluble Ash

1. Add 25 ml of 10% HCl to the dry ash obtained.
2. Cover the crucible with watch glass and heat nearly to boiling for 5 mins.
3. Filter through an ashless filter paper and rinse with hot distilled water several times.
4. Dry and reash filter paper at 500°C.
5. Weigh and calculate as % acid insoluble ash. Then, subtract this from total ash to obtain % acid soluble ash.

*FORMULA USED

1. Total ash percentage $= \dfrac{\text{Weight of ash}}{\text{Weight of sample}} \times 100$

2. % Acid insoluble ash $= \dfrac{\text{Weight of insoluble ash}}{\text{Weight of ash}} \times 100$

3. % Acid soluble ash = 100 – % acid insoluble ash

4. % Water insoluble ash $= \dfrac{\text{Weight of insoluble ash}}{\text{Weight of ash}} \times 100$

5. % Water soluble ash = 100 – % water insoluble ash

OBSERVATIONS

S. No.	Sample	Wt. of crucible	Wt. of sample	Wt. of dried sample	% Ash avg.

Table showing different ash values for experimental samples

S. No.	Sample	% Total Ash	% Acid insoluble ash	% Acid soluble ash	% Water soluble ash	% water insoluble ash

Comparison between experimental values and standard values of ash

Sample	% Ash		Reference
	Experimental value	Standard value	

RESULTS

Ash contents of fresh foods rarely exceed 5%, although some processed foods can have ash contents as high as 12%, e.g., dried beef. Pure oils and fats generally contain little or no ash. The ash content (%) has been found to be __________.

PRECAUTIONS

1. Tarred crucible should be used.
2. Muffle furnace should not be frequently opened or closed.
3. All the weighing should be done carefully.
4. Ashless filter paper should be used.

5.2.2 Determination of the Mineral Content—Calcium, Phosphorus and Iron

INTRODUCTION

The determination of minerals in foods responds to dual concerns. On one hand, it gives the knowledge of the food's nutritional value which refers to the elements called the **major or oligoelements.** On the other hand, it is concerned with verifying if the food contains some minerals in quantities dangerous to the health of the consumers, whether their presence is natural or is due to contamination.

A. CALCIUM ESTIMATION

Calcium is determined by precipitation. It is converted to calcium oxalate precipitate which is dissolved in hot dilute sulfuric acid. It is liberated as oxalic acid which then is titrated with standard potassium permanganate solution.

$$CaCl_2 + (NH_4)_2\,C_2O_4 \rightarrow 2NH_4Cl + CaC_2O_4$$

$$CaC_2O_4 + H_2SO_4 \rightarrow CaSO_4 + H_2C_2O_4$$

$$5H_2C_2O_4 + 2KMnO_4 + 3H_2SO_4 \rightarrow K_2SO_4 + 2MnSO_4 + 8H_2O + 10CO_2$$

MATERIALS REQUIRED

Apparatus: Silica crucible, muffle furnace, weighing balance, water bath, Whatman paper No. 42

Glassware: Watch glass, Beaker, Conical flask, Funnel, Burette, Pipette, Glass rod

Reagents:

1. **Ammonium oxalate:** Saturated solution
2. **Methyl red indicator:** Dissolve 0.5 g methyl red in 100 ml of 95% alcohol
3. **Dilute acetic acid (1:4):** Dissolve 10 ml of acetic acid in 40 ml of distilled water
4. **Dilute NH_4OH (1:4):** Dissolve 10 ml of NH_4OH in 40 ml of distilled water
5. **Dilute H_2SO_4:** Dissolve 20 ml of concentrated H_2SO_4 in 80 ml distilled water. Add acid to water slowly and with constant stirring
6. **0.1 N Potassium permanganate (3.2606 g of $KMnO_4$ per liter)**

   Dissolve 3.3 g of dry potassium permanganate in about 200 ml of water, transfer to a 1-liter flask and make up to volume to 1000 ml. To standardize, weigh out accurately three lots of 0.25-0.30 g of dry sodium oxalate samples having an assay value of 99.95%.

Transfer each portion to a 500 ml beaker, using 250 ml of dilute H_2SO_4 (5 + 95). Stir until the oxalate has dissolved, then add rapidly from a burette about 95% of the amount of permanganate needed for complete oxidation of the sample. Allow the solution to stand until the permanganate is decolorized, then heat to 55-66°C. Complete the titration at this temperature, stirring gently and allowing each drop to become decolorized before adding the next; titrate to first permanent pink color.

$$\text{Normality of } KMnO_4 = \frac{\text{Wt. of sodium oxalate (g)} \times 1000}{\text{ml of } KMnO_4 \times 67.0}$$

where 67 is the Eq. wt. of sodium oxalate.

7. **0.01 N $KMnO_4$ (working solution):** Dilute 10 ml of 0.1 N $KMnO_4$ solution to 100 ml with distilled water (1 ml = 0.2 mg of Ca). Prepare fresh solution before every use.

Sample/s Used: Calcium rich or fortified food samples

PROCEDURE

I. Preparation of the Ash Solution

1. Weigh accurately a suitable amount of sample in tarred silica dish.
2. Volatilize the organic contents by heating just over a low direct flame.
3. Transfer the silica crucible with sample to a muffle furnace (550°C) and heat till the ash is white in color.
4. If the ash is not white, add 1-2 ml of concentrated HNO_3, evaporate to dryness on an indirect flame (on boiling water bath), put back in the muffle furnace for 1-2 hrs. Allow it to cool, note the weight of the ash.
5. Add 40-50 ml of dil HCl(1:1) step by step. Cover the crucible with the help of a watch glass (to prevent spurting). Heat over a water bath for 30 min.
6. Remove the cover. Continue heating for another 30 min to dehydrate the contents.
7. Add another 10 ml of HCl (1:1) to dissolve soluble salts.
8. Filter into 100 ml volumetric flask using Whatman filter paper no. 44. Wash the residue on filter paper with HCl. Make the volume up to 100 ml with distilled water. This is the ash solution for further mineral estimation.
9. Return the filter paper to the crucible, ignite and place it in the muffle furnace at 450°C.
10. Cool and weigh. This gives an approximate amount of silica present in a sample.

II. Determination of the calcium content

1. Pipette 20-100 ml of the aliquot of filtrate, i.e., the ash solution in a 250 ml beaker.
2. Add 25-50 ml of distilled water if necessary. To this, add 10 ml of saturated ammonium oxalate solution and 2 drops of methyl red indicator.
3. Make the solution slightly alkaline by the addition of dilute ammonia and then acetic acid (few drops) until the color is faint pink.
4. Heat the solution to boiling point and allow it to stand for at least 4 hrs or overnight (preferably).
5. Filter with Whatman filter paper no. 42.
6. Wash with distilled water till the filtrate is oxalate free.
7. Since HCl has been used, it is convenient to test for the absence of chloride using $AgNO_3$.
8. Break the precipitate on filter paper with pointed glass rod.
9. Wash the precipitate first using hot dil. H_2SO_4 (1:4) from wash bottle into the beaker in which calcium was precipitated and then with hot water.
10. Titrate while still hot against 0.01 N $KMnO_4$ to first permanent pink color.
11. Finally add filter paper to the solution to complete the titration.

CALCULATIONS

Formula used:

$$\text{Calcium content (mg/100 g)} = \frac{\text{T.V.} \times 0.2 \times \text{V} \times 100}{\text{V} \times \text{W}}$$

where,

T.V. = volume of $KMnO_4$ used for titration
V = total volume of ash solution
v = volume of ash solution taken for estimation
W = weight of sample taken for ashing

OBSERVATIONS

S. No.	*Volume of ash solution (ml)*	*Aliquot of ash solution (ml)*	*Wt. of sample taken for ashing (g)*	*Ca (mg/100 g)*

RESULTS

The amount of calcium was found in the given samples to be__________.

B. PHOSPHORUS ESTIMATION

Phosphorus is converted to a phosphomolybdate complex on reaction with molybdic acid. This complex is then reduced with aminonaphtholsulphonic acid to the complex molybdenum blue which is measured colorimetrically at 650 nm. The intensity of the color is proportional to the phosphate concentration.

Phosphorus + Ammonium molybdate $\xrightarrow{\text{acid pH}}$ Ammonium phosphomolybdate

Ammonium phosphomolybdate + aminonaphthol-sulfonic acid $\rightarrow$ Heteropolybdenum (Blue)

The gravimetric method is used for samples containing more than 10 mg of phosphorus.

MATERIALS REQUIRED

Apparatus: Muffle furnace, Weighing balance, Water bath, Spectrophotometer, Burette stand, Whatman paper no. 42.

Glassware: Silica crucible, watch glass, beaker, conical flask, funnel, burette, pipette, glass rod

Reagents:

1. **Molybdate solution:** Dissolve 25 g of ammonium molybdate in 400 ml of water. Add 500 ml of 10 N H_2SO_4 and make up to 1 liter with distilled water.
2. **Aminonapthol sulphonic acid solution:** Dissolve in distilled water, 0.5 g 1-amino-2-napthol-4-sulphonic acid, 30 g sodium bisulfate ($NaHSO_3$) and 6 g sodium sulphite (Na_2SO_3). Make up to 250 ml. Allow it to stand overnight and filter. Prepare fresh solution every two weeks.
3. **Standard phosphate solution:** Dissolve 0.4389 g of potassium dihydrogen phosphate (KH_2PO_4) in distilled water, add 10 ml of 10 N H_2SO_4 and make up to 1 liter with distilled water.
   (1 ml = 0.1 mg P). Add 1 ml chloroform as preservative.

Sample/s Used: Food sample rich in or fortified with phosphorus

PROCEDURE

1. Take 5 ml of ash solution.
2. Add 5 ml of molybdate reagent solution and mix
3. Add 2 ml of aminonaptholsulphonic acid solution and mix.
4. Make up the volume up to 50 ml and allow it to stand for 10 min.
5. Take absorbance at 650 nm (introducing a blank first).
6. Plot a standard curve and find out the amount of P in the sample.

Preparation of the Standard *P* Series

Dil. no.	*Aliquot from standard solution*	*Amt. of P (1 ml = 0.02 mg)*	*Molybdate solution (ml)*	*Aminonapthol-sulphonic acid solution*	*Distilled water*	*O.D.*
Blank	0	0.0	5	2	43	
T1	5	0.1	5	2	38	
T2	10	0.2	5	2	33	
T3	15	0.3	5	2	28	
T4	20	0.4	5	2	23	
T5	25	0.5	5	2	18	
T6	30	0.6	5	2	13	
T7	35	0.7	5	2	8	
T8	40	0.8	5	2	3	
Sample	–	–	5	2	43	

The standard series will be used to plot a standard curve of absorbance vs concentration of phosphate (ppm) as shown:

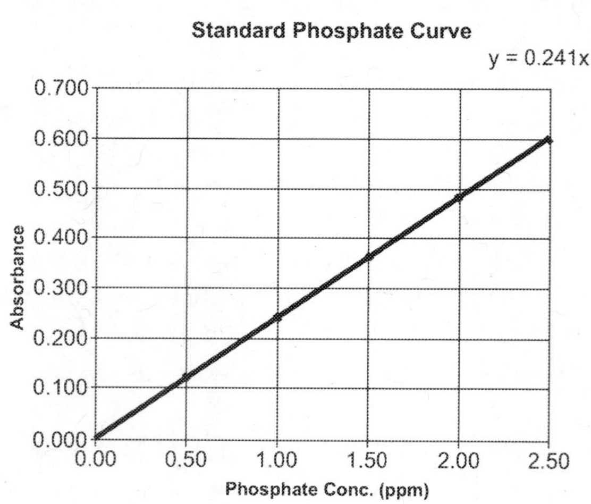

FORMULA USED

$$\text{Phosphorus content (mg/100 g)} = \frac{W \times V \times 100}{V \times W}$$

where,

W = mg of P in aliquot of ash taken for estimation

V = total vol. of ash solution

v = ml of ash solution taken for estimation

w = weight of sample taken for ashing

OBSERVATIONS

S. No.	*Sample*	*Total vol of ash solution (ml)*	*O.D. at 650 nm*	*Aliquot of ash solution (ml)*	*Wt. of sample taken for ashing (g)*	*mg of P reading from std. graph*	*P (mg/100 g)*

RESULTS

The amount of phosphorus in the given sample is __________.

C. IRON ESTIMATION

Iron is determined by converting the iron to ferric form using oxidizing agent - potassium persulfate and treating thereafter with potassium thiocynate to form the red ferric thiocyanate which is measured calorimetrically at 480 nm. The reaction of the iron (III) ion with the thiocyanate ion yields a reddish-brown complex ion shown by the following equation:

$$Fe^{3+}\ (aq) + 6\ SCN^{-}\ (aq) \leftrightarrows [Fe(SCN)_6]^{3-}\ (aq)$$

(Reddish-brown)

MATERIALS REQUIRED

Apparatus: Muffle furnace, Weighing balance, Water bath, Spectrophotometer, Burette stand, Whatman paper no. 42.

Glassware: Silica crucible, Watch glass, Beaker, Conical flask, Funnel, Burette, Pipette, Glass rod

Reagents:

1. **Conc. Sulfuric acid**
2. **Saturated potassium persulfate solution:** Shake 7-8 g of reagent grade iron free potassium persulfate with 100 ml of water in a glass stoppered bottle. The undissolved excess settles to the bottom and compensates for loss by decomposition. Shake briefly before using. Keep the reagent in the refrigerator.
3. **3N potassium thiocyanate solution:** Dissolve 146 g of reagent grade potassium thiocyanate in water and dilute to 500 ml (filter if turbid). Add 20 ml of pure acetone to improve the keeping quality.
4. **Standard iron solution:** Dissolve 0.702 g of reagent grade crystalline ferrous ammonium sulfate in 100 ml of water. Add 5 ml of conc. sulfuric acid, warm slightly and add conc. potassium permanganate solution drop by drop until one drop produces a permanent color. Transfer to a 1 liter volumetric flask, rinse with water and make up to volume. This solution contains 0.1 mg of ferric iron per ml and is stable indefinitely.

Sample/s Used: Iron rich or fortified food samples.

PROCEDURE

1. Use the ash solution prepared by dry ashing for color development as per the following table

		Blank (ml)	*Standard (ml)*	*Sample (ml)*
(i) Standard iron solution (1 ml = 0.1 mg of Fe)	0.0	1.0	0.0	
(ii) Sample ash solution	0.0	0.0	5.0	
(iii) Water		5.0	4.0	0.0
(iv) Conc. H_2SO_4		0.5	0.5	0.5
(v) Potassium persulfate		1.0	1.0	1.0
(vi) Potassium thiocyanate		2.0	2.0	2.0

In each of the above case, make up the volume to 15 ml with water. Measure the color at 480 nm setting the blank at 100% transmission.

USE OF STANDARD CURVE

If it is desired to construct a calibration curve, pipette 0.0, 0.5, 1.0, 1.5, 2.0 and 2.5 ml of standard iron solution to stoppered measuring cylinders. To each add 0.5 ml of conc. H_2SO_4, 1.0 ml of potassium persulfate, 2.0 ml of potassium thiocyanate and make up the volume to 15.0 ml. Measure color at 480 ml. Plot the absorbance against the concentration.

The concentration of iron in the aliquot of the sample can then be read directly from the calibration curve.

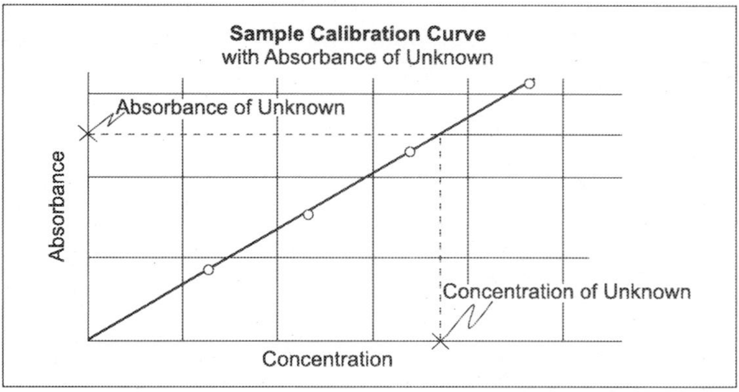

A calibration curve is most useful if the unknown concentration falls within the range of the concentrations used to make the calibration curve. If, however, your unknown has a higher absorbance than your highest concentration point on the calibration curve, you ought to (1) dilute your unknown to get it into the desired absorbance range, or (2) add some higher concentration points to your calibration curve. Similarly, if your unknown concentration falls below the lowest (non-blank) point on your calibration curve, you need to add some lower concentration points to the curve.

FORMULA USED:

$$\text{Iron (mg/100 gm)} = \frac{\text{O.D. of sample} \times 0.1 \times V \times 100}{\text{O.D. of standard} \times 5 \times W}$$

where

V = Total volume of ash solution

W = Weight of sample taken for ash

O.D. = Optical density at 480 nm

RESULTS

The amount of iron in the given sample is __________.

PRECAUTIONS

1. Ash should be kept in muffle furnace till it is completely white.
2. Heating of the ash with acid should be done carefully.
3. Titration for the Ca content should be done with Whatman filter paper No. 42.
4. The O.D. of phosphorus should be taken after 10 min only.
5. Ash solution should not be stored for a long time.
6. Conc. sulfuric acid free from iron should be used for iron estimation.

REVIEW QUESTIONS

1. What is wet ashing?
2. Describe the procedure for dry ashing.
3. What is the role of 1:1 HCl in ashing?
4. Give the principle involved in calcium estimation.
5. You wish to have at least 100 mg of ash from a cereal grain. Assuming 2.5% ash on average, how many grams of the grain should be weighed for ashing?
6. Why is the estimation of minerals critical in food analysis?

5.3 FATS AND OILS

Fats and oils are recognized as essential nutrients in both human and animal diets. Nutritionally, they are concentrated sources of energy (9 cal/gram); provide essential fatty acids which are the building blocks for the hormones needed to regulate bodily systems; and are a carrier for the oil soluble vitamins A, D, E, and K. They also enhance the foods we eat by providing texture and mouth feel, imparting flavor, and contributing to the feeling of satiety after eating. Fats and oils are also important functionally in the preparation of many food products. They act as tenderizing agents, facilitate aeration, carry flavors and colors and provide a heating medium for food preparation.

Fats and oils are classified as "lipids" which are usually defined as those components that are soluble in organic solvents (such as ether, hexane or chloroform), but are insoluble in water.

This group of substances includes triacylglycerols, diacylglycerols, monoacylglycerols, free fatty acids, phospholipids, sterols, caretonoids and vitamins A and D. The lipid fraction of a fatty food therefore contains a complex mixture of different types of molecule. Even so, triacylglycerols are the major component of most foods, typically making up more than 95 to 99% of the total lipids present. Triacylglycerols are esters of three fatty acids and a glycerol molecule. The fatty acids normally found in foods vary in chain length, degree of unsaturation and position on the glycerol molecule. Consequently, the triacylglycerol fraction itself consists of a complex mixture of different types of molecules. Each type of fat has a different profile of lipids present which determines the precise nature of its nutritional and physicochemical properties.

They have lower densities than water and may have consistencies at ambient temperature of solid, semisolid, or clear liquid. When they are solid appearing at a normal room temperature, they are referred to as **"fats,"** and when they are liquid at that temperature, they are called **"oils."**

Many fats especially unsaturated ones are prone to lipid oxidation, which leads to the formation of off-flavors and potentially harmful products. Some of the most important properties of concern to the food analyst are:

- Total lipid concentration
- Type of lipids present
- Physicochemical properties of lipids, e.g., crystallization, melting point, smoke point, rheology, density and color
- Structural organization of lipids within a food

Diagrams of simple and mixed triglycerides

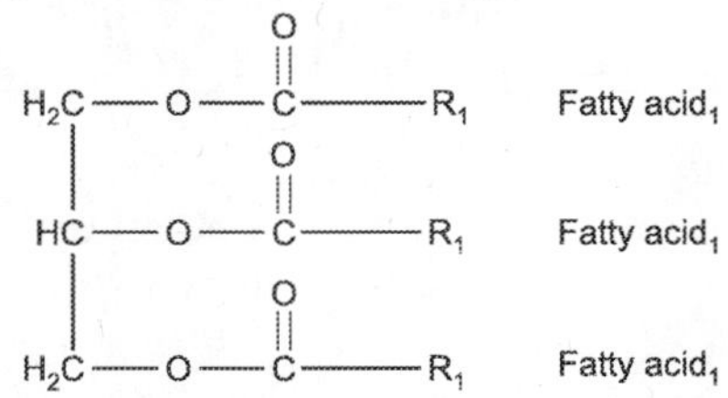

Simple Triglyceride

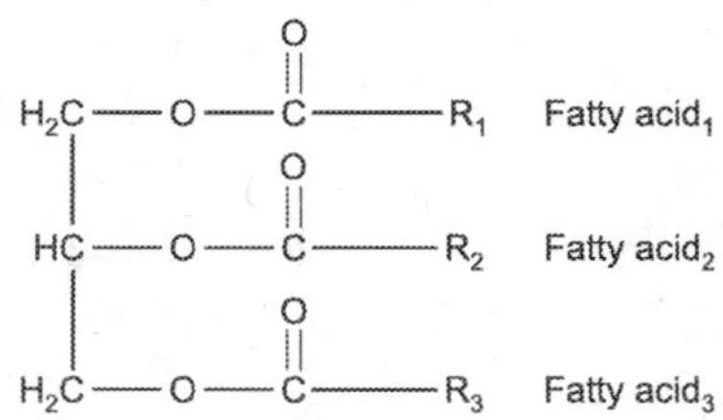

Mixed Triglyceride

A number of chemical methods have been developed to provide information about the type of lipids present in edible fats and oils. These techniques are much cruder than chromatography techniques because they only give information about the average properties of the lipid components present, e.g., the average molecular weight, degree of unsaturation or amount of acids present. Nevertheless, they are simple to perform and do not require expensive apparatus and so, they are widely used in industry and research.

Although knowledge of the total amount of fat in a food is important, information regarding the nature of the fats or oils, particularly the degree of saturation, is also of paramount importance, since the nature of the fat or oil may affect keeping qualities, texture and nutritional value. Deterioration of fatty foods may involve either hydrolysis (lipolysis) of the fat by lipase enzymes, releasing free fatty acids and resulting in hydrolytic, or lipolytic, rancidity, or oxidation of fats, culminating eventually in oxidative rancidity as a result of the production of compounds of undesirable flavor.

Lipolysis is of most significance in dairy products, since milk fat differs from most other fats in having a high level of volatile fatty acids, such as butyric acid, which produce particularly undesirable flavors if released. Oxidative rancidity is of greater importance with fats and oils of high unsaturation since the process of oxidation occurs more readily with increasing unsaturation. Thus, fish oils are most susceptible, polyunsaturated vegetable oils very susceptible and animal fats least.

Generally, fats and oils are tested to gain information relating to performance in specific food applications, to measure degree of deterioration (such as oxidation or rancidity) as well as stability of fat against such change, to check fat properties against purchase specifications, and to identify fats and oils against possible misrepresentation or adulteration. The important physical and chemical properties of fats and oils have been measured in various ways that have created a number of terms commonly associated with fats.

There is a need for the analysis of the chemical properties of fat and oil because issues such as the effect of dietary fat on health and food labeling requirements necessitate that food scientists be able not only to measure the total lipid content of a food stuff but also to characterize it. Health concerns require the measurement of parameters such as cholesterol content, amount of saturated and unsaturated fat, and perhaps even the type and amount of individual fatty acids. Measurements of lipid stability impact not only the shelf life of the product but also its safety.

5.3.1 Determination of Moisture Content by Air-Oven Method

INTRODUCTION

Oils and fats are important parts of human diet and more than 90 per cent of the world production from vegetable, animal and marine sources is used as food or as an ingredient in food products. Oils and fats are a rich source of dietary energy and contain more than twice the caloric value of equivalent amount of sugar. Their functional and textural characteristics contribute to the flavor and palatability of natural and prepared foods. They contain certain fatty acids which play an important role in nutrition and are also carriers of fat soluble vitamins. Moisture content of oils and fats is the loss in mass of the sample on heating at 105°C under the operating conditions specified.

MATERIALS REQUIRED

Apparatus: Aluminum dishes 7 – 8 cm in diameter and 2 - 3 cm deep provided with tight fitting lids as covers, Hot air oven, Weighing balance

Sample/s used: Fat and oil samples

PROCEDURE

I. Preparation of Test Sample

Liquid oils

1. Use clear sediment free liquid directly after inverting container several times.
2. If liquid sample contains sediment, release all sediment from walls of container and distribute uniformly throughout the oil for determination of moisture.

Solid and Semisolid Samples

1. Soften the sample if necessary by gentle heat, taking care not to melt it. When soft enough, mix thoroughly for determination of moisture and volatile matter.
2. For other determinations, melt in drying oven at a temperature at least 10°C above the melting point. If clear, proceed directly. If turbid or contains sediment, filter the test sample.

II. Estimation of Moisture

1. Weigh in a previously dried and tarred dish about 5-10 g of oil or fat which has been thoroughly mixed by stirring.
2. Loosen the lid of the dish and heat it in an oven at 105°C for 1 hour. Remove the dish from the oven and close the lid.
3. Cool in a desiccator containing phosphorus pentoxide or equivalent dessicant and weigh.

4. Heat in the oven for a further period of 1 hour, cool and weigh.
5. Repeat this process until change in weight between two successive observations does not exceed 1 mg. Carry out the determination in duplicate.

CALCULATIONS

Moisture and volatile matter (% by weight) $= \dfrac{W_1 \times 100}{W}$

where,

W_1 = Loss in gm of the material on drying

W = Weight in gm of the material taken for test

RESULTS

The % moisture in the given sample is ________.

Presence of moisture is detrimental to the quality of fats and oils as it induces hydrolysis which leads to a resultant increase in free fatty acids and off flavors.

PRECAUTION

1. Keep the oils and fats in a cool place and protect from light and air to retard rancidity.

5.3.2 Determination of Specific Gravity

INTRODUCTION

The physical properties of the natural fats and oils are often used to identify them. Specific gravity is one such property. Specific gravity (S.G.) of an oil is defined as the ratio of the density of that oil at a particular temperature and the density of water at that same temperature. Specific gravity usually means relative density with respect to water.

The temperature is carefully controlled during determination of specific gravity since, significant changes in the fats and oils occur in short range of temperature. The specific gravity of a fat or oil is usually measured at 25°C, but it may be necessary to use temperatures of 40°C or even 60°C for high-melting fats. In general, either unsaturation of the fatty acid chains or increase in chain length of the fatty acid residues tends to increase the specific gravity.

The specific gravity of an oil can be measured commonly using a pycnometer. Other methods include use of hydrometer floatation method and use of specific gravity centrifuge tube.

FORMULA USED:

$$\text{Specific Gravity} = \frac{\text{Wt. of S.G. bottle and oil sample} - \text{Wt. of empty S.G. bottle}}{\text{Wt. of S.G. bottle and water} - \text{Wt. of empty S.G. bottle}}$$

where S.G. bottle means pycnometer

MATERIALS REQUIRED

Apparatus: Drying oven, Heating mantle, Weighing balance, Water bath, Thermometer, Beaker and Pycnometer

Pycnometer: It is a small, light flask, usually made of glass of definite volume. It can be used to weigh equal volume of liquid at a given fixed

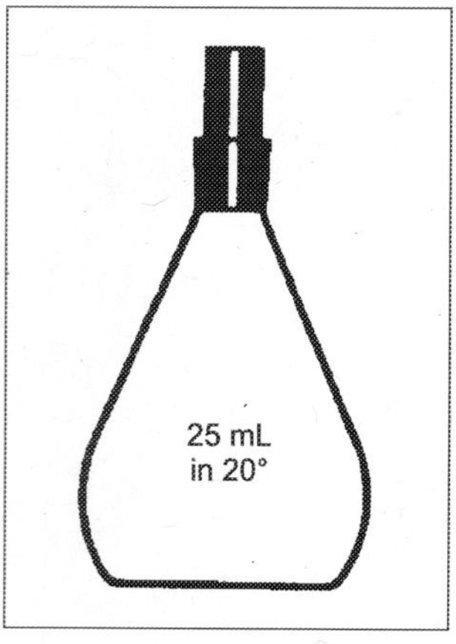

Pycnometer

temperature. The ratio of the weights of these equal volumes yields specific gravity. The pycnometer method is a very accurate means of determining specific gravity.

Sample/s Used: Oil samples such as soybean, groundnut, cocunut, etc.

PROCEDURE

1. Note the tare weights of clean dry pycnometers (one for water and others as per the number of oil samples).
2. Fill one of the pycnometers with recently boiled and cooled distilled water at 20-23°C, insert the stopper and incubate in a water bath at 25° ± 0.2°C for 30 minutes.
3. Remove the bottle from the bath, wipe dry and weigh. Note the weight of the water.
4. Now, fill the other pycnometers with the oil samples at 20-23°C to overflowing.
5. Insert the stopper and repeat the same procedure as for water.
6. Note the weight of oil.

OBSERVATIONS AND CALCULATIONS

S. No.	*Sample*	*Wt. of empty pycnometer (g)*	*Wt. of pycnometer + oil (g)*	*Wt. of oil (g)*	*Specific Gravity* (at 25°C)*

*Formula Used:

$$\text{Specific Gravity} = \frac{\text{Wt. of S.G. bottle and oil sample} - \text{Wt. of empty S.G. bottle}}{\text{Wt. of S.G. bottle and water} - \text{Wt. of empty S.G. bottle}}$$

RESULTS

S. No.	*Sample*	*Specific Gravity*	
		Experimental value	*Standard value**

PRECAUTIONS

1. The glassware used should be thoroughly clean and dry.
2. The temperature of the water bath should be properly maintained.
3. While filling the oil in the pycnometer, no air bubbles should be introduced.
4. The pycnometer should be filled to the top before putting the cork stopper.

5.3.3 Determination of Refractive Index

INTRODUCTION

The index of refraction is the degree of deflection of a beam of light that occurs when it passes from one transparent medium to another. The refractive indices of fats and oils are often measured because they can be rapidly and accurately determined and secondly, they are useful in identification of these substances and the testing of their purity. An Abbe's refractometer with temperature control is used and the measurement is usually at 40°C. The index of refraction decreases as the temperature rises however, it increases with increase in the length of the carbon chains and also with the number of double bonds present.

For Refractive Index

$R = R' + K(T' - T)$

where

R' = observed reading at T′

K = 0.00365 (fats)

K = 0.00385 (oil)

T = Temperature at which refractive index is to be calculated

T = 40°C

MATERIALS REQUIRED

Apparatus: Drying oven, Heating mantle, Weighing balance, Water bath, Thermometer, Beaker

- **Abbe's Refractometer**

Principle of Abbe's Refractometer

Abbe's refractometer works on the principle that a ray of light passing through the diffused surface enters into the liquid mechanism at different angles of incidence. The eyepiece is fixed and the screw is rotated as to get the coincidence of the critical ray with the cross wire of the eyepiece. The setting of the prism at this portion corresponds to a value of refractive index which is read directly on the scale engraved on the instrument.

Sample Used: Oil samples

PROCEDURE

For measurement of refractive index using Abbe's refractometer

1. Adjust the temperature of the refractometer to 40° ± 1°C by circulating water from a constant temperature water bath.
2. Shift the lower mirror to left hand side.
3. Unlock the second mirror and wipe it thoroughly with water and alcohol.

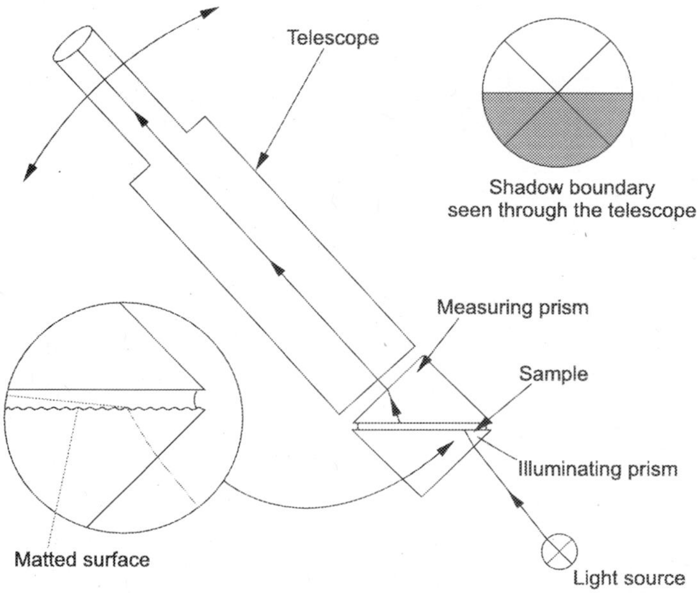

Schematic diagram showing the working of Abbe's refractometer

4. Melt the sample, if not already a liquid and filter through a filter paper.
5. Smear the sample on the cleansed prism and read the refractive index.
6. After taking the reading, clean the prism using luke warm water and wipe dry.
7. Repeat the above procedure for all the samples.

Note: The instrument is calibrated at certain temperature, so to find out the reading at a particular temperature, temperature correction is to be done as shown.

CALCULATIONS

For refractive index

$$R = R' + K(T' - T)$$

where

R' = observed reading at T′

K = 0.00365 (fats)

K = 0.00385 (oil)

T = Temperature at which refractive index is to be calculated

T = 40°C

RESULTS

The refractive index of the given samples are as follows:

S. No.	*Sample*	*Refractive index*	
		Experimental value	*Standard value**

PRECAUTIONS

1. The glassware used should be thoroughly clean and dry.
2. Reading of the refractometer should be noted carefully.
3. Temperature correction should be used in the calculation of the refractive index (if required).

5.3.4 Determination of Smoke Point and Percent Fat Absorption

INTRODUCTION

Smoke point is the temperature at which a thin, whitish blue smoke appears on the surface of oil or fat. This is measured by a standard method using an open dish so that presence of smoke can be noticed.

Smoke point is followed by flash point, i.e., the temperature at which a flash appears at any point on the surface of the sample. Finally, it is the fire point at which evolution of volatiles (by decomposition of samples) proceeds with enough speed to support continuous combustion.

These tests give a measure of the effect of heating on the physicochemical properties of lipids. They are particularly important for selecting lipids that are going to be used at high temperatures, e.g., for frying. The tests reflect the amount of volatile organic material in oils and fats such as free fatty acids.

The smoke point of a fat is partly a matter of its natural composition and partly, a matter of the processing, it has received. Soybean, cottonseed, peanut and corn oils have smoke points of about 230°C. Hydrogenated fats smoke at 221°-232°C. Shortenings containing monoglycerides as an emulisifier, smoke at a lower temperature of about 176°C.

Because fats differ in their smoke points, fat to be used for frying should be chosen on the basis of their resistance to smoking at the temperature used. Smoke point is used to determine purity and adulteration of fats and oils. It also indicates whether the sample has been used earlier for processing. Following factors affect the smoke point:

1. **Percent of free fatty acids:** The development of free fatty acids by hydrolysis of the fat during frying causes a decrease in smoke point. A fat that has had repeated or prolonged use will begin to smoke at a temperature too low for frying.
2. **Suspended matter** such as flour or batter particles also lower the smoke point.
3. **Greater the surface** of the fat exposed, lower is the smoke point.
4. **Length of carbon chain:** Increase in length of carbon chain leads to increase in smoke point.

Table showing description, uses, type and smoke point of different fats and oils.

Fats or oils	*Description*	*Cooking uses*	*Type of fat*	*Smoke point °C*
Butter	Whole butter is a mix of fats, milk solids, and moisture derived by churning cream until the oil droplets stick together and can be separated out.	Baking, cooking	Saturated	177°C
Butter (Ghee), clarified	Ghee has a higher smoke point than butter since clarification eliminates the milk solids (which burn at lower temps).	Frying, sautéing	Saturated	190-250°C (depending on purity),
Coconut Oil	A heavy nearly colorless oil extracted from fresh coconuts.	Coatings, confectionery, shortening	Saturated	177°C
Corn Oil	A mild, medium-yellow color refined oil. Made from the germ of the corn kernel.	Frying, salad dressings, shortening	Polyunsaturated	232°C
Olive Oil	Oils vary in weight and may be pale yellow	Cooking, salad dressings,	Monounsaturated	160°C 216°C 238°C
	to deep green depending on fruit used and processing.	Sauté, pan fry, sear, deep fry, stir fry, grill, baking		242°C
Palm Oil	A yellowish-orange fatty oil obtained especially from the crushed nuts of an African palm.	Cooking, flavoring	Saturated	230°C

Peanut Oil	Pale yellow refined oil with a very subtle scent and flavor. Made from pressed steam-cooked peanuts. Used primarily in Asian cooking.	Frying, cooking, salad dressings	Monounsaturated	232°C
Rice Bran Oil	Rice bran oil is produced from the rice bran, which is removed from the grain of rice as it is processed.	Frying, sauté, salad dressings, baking, dipping oils	Monounsaturated	254°C
Safflower Oil	A golden color with a light texture. Made from the seeds of safflowers.	Margarine, mayonnaise, salad dressings	Polyunsaturated	232°C
Sesame Oil	Comes in two types – a light, very mild Middle Eastern type and a darker Asian type pressed from toasted sesame seeds.	Cooking, salad dressings	Polyunsaturated	232°C
Soybean Oil	A fairly heavy oil with a pronounced flavor and aroma.	Margarine, salad dressings, shortening	Polyunsaturated	232°C
Sunflower Oil	A light odorless and nearly flavorless oil pressed from sunflower seeds. Pale yellow.	Cooking, margarine, salad dressings, shortening	Polyunsaturated	232°C

Contd...

Contd...

Vegetable Oil*	Made by blending several different refined oils. Designed to have a mild flavor and a high smoke point.	Cooking, salad dressings	Polyunsaturated	

Source: The Culinary Institute of America (1996). The New Professional Chef, 6th edition, John Wiley and Sons

*The smoke point of a vegetable oil is variable as it depends on the type of oils/oilseeds used for blending and their percentage.

Fat Absorption

From the standpoint of both palatability and ease or rapidity of digestion, it is desirable to hold fat absorption by fried food to a minimum. Generally, some 10% oil or fat is absorbed but larger amount are absorbed in some conditions making the product become soggy and undesirable.

Fat absorption is the percentage of the fat/oil absorbed by the sample during the frying process.

$$\% \text{ Fat absorption} = \frac{W_2 - W_1}{W_1} \times 100$$

where W_1 = Weight of the sample before frying

W_2 = Weight of the sample after frying

Factors affecting fat absorption are the amount of the food, condition of the frying fat (including the level of surfactants present), temperature of fat and the length of heating time.

MATERIALS REQUIRED

Apparatus: Open pan, Heat source, Oil thermometer, Weighing balance

Sample/s Used: Different types of vegetable oils, fryums

PROCEDURE

I. Smoke point determination

1. Take 250 ml of given sample in a clean open pan for frying.
2. Put the pan over medium flame.
3. Note the temperature at which a thin, whitish blue smoke leaves the pan, using oil thermometer.

II. % Fat absorption determination

1. Weigh 5 gm of fryums at three places.

2. Fry each sample at the smoke point and temperature ± 10°C of the same (e.g., if smoke point is 180°C, then fry at 170°C and 190°C).
3. Note the weight before and after frying, the contact time and the appearance factors (such as color, greasiness or crispiness) and calculate percent fat absorption using the given formula.

II. Effect of frying on smoke point

1. Allow the used oil to cool.
2. Again note the smoke point.

OBSERVATIONS AND CALCULATIONS

Table showing the smoke point and percent fat absorption of given samples of fats and oils

S. No.	Sample	Smoke Point (°C)	Frying Temp. (°C)	W_1 (g)	W_2 (g)	% Fat Abs.*	Contact Time (Sec.)	Appearance		
								Color	Greas- -iness	Crisp- -iness

*Formula Used:

$$\% \text{ fat absorption} = \frac{W_2 - W_1}{W_1} \times 100$$

where W_1 = Weight of the sample before frying
W_2 = Weight of the sample after frying

Table showing effect of reheating on smoke point

S. No.	Oil sample	Smoke point °C	
		Fresh sample	Used sample

RESULTS

S. No.	Sample	Smoke point °C	Fat absorption (%) at smoke point

The more refined an oil, the higher is the smoke point. The reason being that refining removes the impurities that can cause the oil to smoke. Different fats and oils have different uses. Each performs best within a certain range of temperature. Some are made for high heat cooking, while others have intense flavors that are best enjoyed by drizzling directly on food.

Knowing the smoke point of an oil is important because heating oil to the point where the oil begins to smoke produces toxic fumes and harmful free radicals. The smoke point of oil depends to a very large extent on its purity and age at the time of measurement. A simple rule of thumb is that the lighter the color of the oil, the higher its smoke point. When frying, it is important to choose an oil with a very high smoking point. Most foods are fried between the temperatures of 177-233°C so it is best to choose an oil with a smoking point above 205°C.

The chief factors governing the amount of fat which will be absorbed by a food during frying are:

1. The time of cooking
2. The temperature of the cooking fat
3. The total surface area of the food
4. The composition and nature of the food

Time of cooking: In general, it can be stated that the longer the food is cooked, the greater the fat absorption. There will be some exceptions to this. For instance, with some foods cooked at high temperatures, coagulated material or a hardened crust may prevent greater fat absorption with longer cooking. Other foods which contain a high percentage of fat like meat may lose fat during cooking. This is also true for fried chicken, if the chicken has a high fat content.

Temperature of cooking fat: Usually, there is less fat absorption at higher cooking temperatures. This is probably due to the quick formation of a crust or coagulated material at the high temperature. However, slight differences in cooking temperature, 30° to 40°C, cause little variation in fat absorption. Ordinarily, food is cooked at moderate temperatures to thoroughly cook the food without excessive browning.

Surface area: The larger surface for a given weight of material, the greater the area over which fat may be absorbed. Cracks caused by handling or expansion in cooking, rolling thin rather than thick, roughened surfaces due to little mixing or other causes and stretching increase the surface area and thus tend to increase the fat absorption.

Composition and nature of the food: Some foods by their nature tend to absorb more or less fat than others, e.g., fried eggs may absorb less fat than some foods. This is due to the smooth surface and the coagulation of the egg proteins, which tends to prevent fat absorption. Large amounts of flour on the surface of cutlets, pakoras may tend to increase the fat absorption, owing to the absorption of fat by the flour. Variations in composition of the food and variations in the proportion of ingredients used in foods cause wide variations in the amount of fat absorbed.

PRECAUTIONS

1. The thermometer should not touch the base of the container.
2. Temperature should be noted carefully.
3. The contact time should be noted from putting the sample into the oil till complete frying is over.
4. Care should be taken so as not to char the sample.
5. Initial and final weight of the sample should be taken on the same piece of butter paper.

5.3.5 Determination of Melting Point of Fat by Open-Tube Capillary-Slip Method

INTRODUCTION

Oils and fats are chiefly mixtures of triglycerides. They do not exhibit either a definite or sharp melting point. Therefore, the melting point does not imply the same characteristics that it does with pure crystalline substances. Fats pass through a stage of gradual softening before they become completely liquid. The melting point is therefore defined by the specific conditions of the method by which it is determined.

Open-Tube Capillary-Slip Method

Principle: The temperature at which the oil or fat softens or becomes sufficiently fluid to slip or run as determined by the open-tube capillary-slip method.

MATERIALS REQUIRED

Apparatus: Heat source, Thermometer with 0.2°C sub-divisions and a suitable range.

Glassware:

1. Melting point tubes

   Thin walled with uniform capillary glass tubes open at both ends with the following dimensions:

   (i) Length 50 to 60 mm

   (ii) Inside dia. 0.8 to 1.1 mm

   (iii) Outside dia. 1.2 to 1.5 mm

2. Beaker with a side tube heating arrangement. (Thiele melting point tube may be used).

Sample/s Used: Butter, pure ghee

PROCEDURE

1. Melt the sample and filter it through a filter paper to remove any impurities and last traces of moisture. Make sure that the sample is absolutely dry.
2. Mix the sample thoroughly. Introduce a capillary tube into the molten sample so that a column of the sample, about 10 mm long, is sucked into the tube.
3. Chill the tube containing the sample immediately by touching the tube containing ice until the fat solidifies.

4. Place the tube in a small beaker and hold it for one hour either in a refrigerator or in water maintained at 4°C to 10°C.
5. Remove the tube and attach it, with a rubber band, to the thermometer bulb so that the lower end of the capillary tube and the thermometer bulb are at the same level.
6. Take water at 10°C in the 'Thiele' tube and immerse the thermometer with the capillary tube containing the sample of fat. Gradually increase the temperature by heating at the side-tube of the Thiele tube at the rate of 2°C per min, till the temperature reaches 25°C, and thereafter at the rate of 0.5°C per min.
7. Note the temperature of the water when the sample column begins to rise in the capillary tube. Report the average of two such separate determinations as the melting point, provided that the readings do not differ by more than 0.5°C.

RESULTS

The melting point of the given sample is________.

The degree of saturation or unsaturation determines not only a fat's melting point, but also its chemical stability and its likelihood of auto-oxidizing and creating harmful free radicals. Unsaturated oils melt at lower temperatures and the more unsaturated they are, the lower the temperature at which they are viable.

PRECAUTIONS

1. The capillary tube should contain the fat column about 10 mm long.
2. Note the temperature carefully when the sample column begins to rise.

5.3.6 Determination of Saponification Value

INTRODUCTION

Saponification is the process of breaking down or degrading a neutral fat into glycerol and fatty acids by treatment of the fat with alkali. The saponification value of an oil or fat is defined as the number of milligrams of alkali (KOH) required to neutralize the fatty acids resulting from the complete hydrolysis of 1 g of the sample.

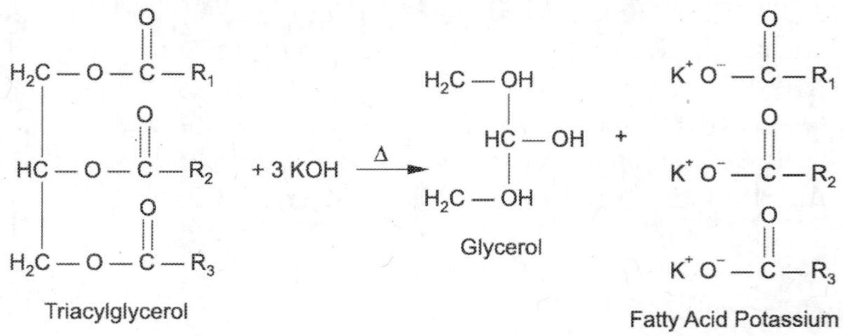

For example:

$$C_3H_5(C_{17}H_{35}COO)_3 + 3KOH \xrightarrow{\Delta} C_3H_5(OH)_3 + 3C_{17}H_{35}.COOK$$

Stearin   Glycerol   Potassium stearate

The esters of the fatty acids of low molecular weight require the most alkali for saponification, so that saponification value (S.V.) is inversely proportional to the mean of the molecular weight of fatty acids in the glycerides present.

Application: The saponification number is an index of the average mol. wt. of the triacylglycerols in the samples. The mol. wt. of the triacylglycerols may be divided by 3 to give an approx. mol. wt. for the fatty acids present. The smaller the saponification number, the longer the average fatty acid chain length.

$$\text{Average molecular weight} = \frac{56.1}{SN} \times \frac{1000\ \text{mg}}{\text{g}} \times 3$$

where 56.1 = M.W. of KOH

SN = Saponification number in mg/g.

The S.V. is of some use for detecting the presence of coconut oil (SV248-265), palm kernel oil (SV-230-254) and butter fat (SV225), which contain a high proportion of the lower fatty acids.

The presence of any paraffin hydrocarbons may be detected as a cloudiness when water is added to the ethanol saponified oil or fat solution.

MATERIALS REQUIRED

Apparatus: Heating mantle, Burette stands, Sealing wax, Weighing balance

Glassware: Distillation flask (250 ml capacity and having a ground glass joint), Pipette, Glass beads, Air condenser (202 cm long), Burette, Beaker, Measuring cylinder

Reagents:

1. **0.5 N Alcoholic KOH:** Dissolve 30 g of KOH in 20 ml of distilled water and add sufficient alcohol to make the volume up to 1000 ml. Allow to stand overnight, decant the clear liquid and store in a bottle with rubber stopper.
2. **0.5 N HCl**
3. **0.5 N NaOH**
4. **Neutralized alcohol**
5. **Phenolphthalein**

Sample/s Used: Vegetable oil samples

PROCEDURE

1. Weigh 1-2 gm of sample in the distillation flask.
2. To this add 25 ml of alcoholic KOH and a few glass beads.
3. Connect the flask with air condenser and reflux it for less than one hour at 25-30°C on a heating mantle till the sample gets clear.
4. Allow the flask and condenser to cool to room temperature.
5. Add 10 ml of lukewarm, neutralized alcohol through the condenser to wash away adhering particles.
6. Remove the condenser and add 1 ml of phenolphthalein indicator to the contents of the flask and titrate against 0.5 N HCl. Simultaneously, also conduct a blank without the sample.

*FORMULA USED

$$\text{S.V.} = \frac{(\text{Blank T.V.} - \text{Sample T.V.}) \times \text{Normality of HCl used} \times 56.1}{\text{Weight of sample (g)}}$$

where 56.1 = M.W. of KOH

OBSERVATIONS AND CALCULATIONS

1. Sample weight (calculate using differential weight method)
2. Titration of blank v/s 0.5 N HCl

S. No.	Initial reading (ml)	Final reading (ml)	Volume used (ml)

3. Titration of sample v/s 0.5 N HCl

S. No.	Initial reading (ml)	Final reading (ml)	Volume used (ml)

*Formula Used

$$\text{S.V.} = \frac{(\text{Blank T.V.} - \text{Sample T.V.}) \times \text{Normality of HCl used} \times 56.1}{\text{Weight of sample (g)}}$$

where 56.1 = M.W. of KOH

RESULTS

Table showing saponification value for various experimental samples.

S. No.	Sample	Wt. of sample (g)	Titer value for sample (ml)	Titer value for blank (ml)	Saponification value

Table showing comparison of experimental results of SV with standard reference values.

S. No.	Sample	Saponification Value mg KOH/g oil		Reference*
		Experimental value	Theoretical value	

Refer E at page 159

PRECAUTIONS

1. The glassware must be washed thoroughly as any traces of detergent may contribute to saponification.
2. The refluxed sample must be cooled before further use.
3. Neutralised alcohol should be lukewarm.

5.3.7 Determination of Percent Free Fatty Acids and Acid Value

INTRODUCTION

Fats are degraded by the process of hydrolysis, which splits triglycerides into their basic components of glycerol and free fatty acids in the presence of moisture. The free fatty acids, especially if they are of short-chain cause off-odors and rancid flavors in fats and oils. The term **acid value** refers to a measure of free fatty acids present in a fat.

Measures of fat acidity normally reflect the amount of fatty acids hydrolyzed from triacylglycerols. In addition to free fatty acids, acid phosphates and amino acids can also contribute to acidity. Acid value is defined as the mg of KOH necessary to neutralize the free acids present in 1 g of fat or oil. Free fatty acid is the percentage by weight of a specified fatty acid (e.g. % oleic acid).

In samples containing no acids other than fatty acids, acid value and FFA may be converted from one to the other using a conversion factor. Sometimes the acidity of edible oils and fat is expressed as ml of NaOH required to neutralize the fatty acids in 100 g of fat or oils.

Application: In crude fat, acid value of free fatty acids estimates the amount of oil that will be lost during refining steps designed to remove fatty acids. In refined fats, a high acidity level means a poorly refined fat or fat breakdown after storage or use. If the fatty acids liberated are volatile, acid value or FFA may be a measure of hydrolytic rancidity. Fatty acids oxidize at a slightly higher rate when free than when esterified to glycerol. It is an indicator of faulty processing or storage. It tells about quality of fat or oil and its shelf life. It indicates whether a sample is fresh or used.

MATERIALS REQUIRED

Apparatus: Heat source, pan, Weighing balance

Glassware: Conical flask, Measuring cylinder, Beaker, Burette

Reagents:

1. **Neutralized alcohol:** Add few drops of phenolphthalein to the required volume of ethanol followed by few drops of 0.1 N NaOH till the color changes to light pink.
2. **0.1 N NaOH**
3. **Phenolphthalein indicator.**

Sample/s Used: Fresh cooking oils

PROCEDURE

1. Weigh 5 g of a given sample in a dry and clean conical flask.
2. Melt the sample gently in a warm water bath.

3. Add 50 ml of neutralized alcohol.
4. Titrate against 0.1 N NaOH using phenophthalein as an indicator under warm conditions and continuous stirring.
5. End point is the appearance of faint pink color.
6. Repeat twice for concordant titer values.

FORMULA USED

For NaOH and predominant acid (oleic acid)

$$\text{\% FFA (as oleic acid)} = \frac{\text{Titer value} \times \text{N of NaOH} \times 28.2}{\text{Weight of sample}}$$

$$\text{Acid value (AV)} = \frac{\text{Titer value} \times \text{Normally of NaOH} \times 40}{\text{Weight of sample}}$$

For KOH and predominant acid (oleic acid)

$$\text{\% FFA (as Oleic acid)} = \frac{\text{Titer value} \times \text{N of KOH} \times 28.2}{\text{Weight of sample}}$$

$$\text{Acid Value (AV)} = \frac{\text{Titer value} \times \text{Normally of KOH} \times 56.1}{\text{Weight of sample}}$$

OBSERVATIONS

Table showing volume of HCl used for neutralization of FFA

S. No.	*Sample*	*Weight of Sample (g)*	*Titer value (ml)*		
			Initial	*Final*	*Volume used*

RESULTS

S. No.	*Sample*	*Type*	*Experimental values*		*Reference values**	
			% FFA	*Acid value*	*% FFA*	*Acid value*

Refer E at page 159

Evaluation of FFA content is an important aspect of judging the quality of any fat or oil sample. The free fatty acids, especially when they are short chains, cause off-odors and rancid flavors.

PRECAUTIONS

1. Weighing should be accurate.
2. Alcohol should be neutralized prior to use.
3. Titration should be carried out in warm conditions.
4. %FFA should be calculated as per the predominant fatty acid present.

5.3.8 Determination of Iodine Value using Wiz's Method

INTRODUCTION

The iodine value (I.V.) gives a measure of the average *degree of unsaturation* of a lipid: the higher the iodine value, the greater the number of C=C double bonds. By definition, the iodine value is expressed as the grams of iodine absorbed per 100 g of lipid. One of the most commonly used methods for determining the iodine value of lipids is "Wiz's method". The lipid to be analyzed is weighed and dissolved in a suitable organic solvent, to which a known excess of iodine chloride is added. Some of the ICl reacts with the double bonds in the unsaturated lipids, while the rest remains:

$$R-CH=CH-R + ICl_{excess} \rightarrow R\text{-}CHI-CHCl-R + ICl_{remaining}$$

The amount of ICl that has reacted is determined by measuring the amount of ICl remaining after the reaction has gone to completion ($ICl_{reacted} = ICl_{excess} - ICl_{remaining}$). The amount of ICl remaining is determined by adding excess potassium iodide to the solution to liberate iodine, and then titrating with a sodium thiosulfate ($Na_2S_2O_3$) solution in the presence of starch to determine the concentration of iodine released:

$$ICl_{remaining} + 2\ KI \rightarrow KCl + KI + I_2$$

$$I_2 + \text{Starch} + 2\ Na_2S_2O_3\ (\text{blue}) \rightarrow 2\ NaI + \text{Starch} + Na_2S_4O_6\ (\text{colorless})$$

Iodine itself has a reddish brown color, but this is often not intense enough to be used as a good indication of the end point of the reaction. For this reason, starch is usually used as an indicator because it forms a molecular complex with the iodine that has a deep blue color. Initially, starch is added to the solution that contains the iodine and the solution goes a dark blue. Then, the solution is titrated with a sodium thiosulfate solution of known molarity. While if there is any I_2 remaining in the solution, it stays blue, but once all of the I_2 has been converted to I^- it turns colorless. Thus, a change in solution appearance from blue to colorless can be used as the end point of the titration.

The concentration of C=C in the original sample can therefore be calculated by measuring the amount of sodium thiosulfate needed to complete the titration. The higher the degree of unsaturation, the more iodine absorbed, and the higher the iodine value.

The iodine value is used to obtain a measure of the average degree of unsaturation of oils, and to follow processes such as hydrogenation and oxidation that involve changes in the degree of unsaturation.

FORMULA USED

$$\text{Iodine value} = \frac{(\text{Titer value B} - \text{Titer value S}) \times N \times 12.69}{W}$$

S = Sample used

B = Blank

N = Normality of Thiosulfate sol.

W = Weight of the sample taken

12.69 is used to convert from meq. thiosulfate to g iodine; M.W. of iodine is 126.9.

MATERIALS REQUIRED

Apparatus: Weighing balance

Glassware: Iodine flask, Pipette, Measuring cylinder and Burette

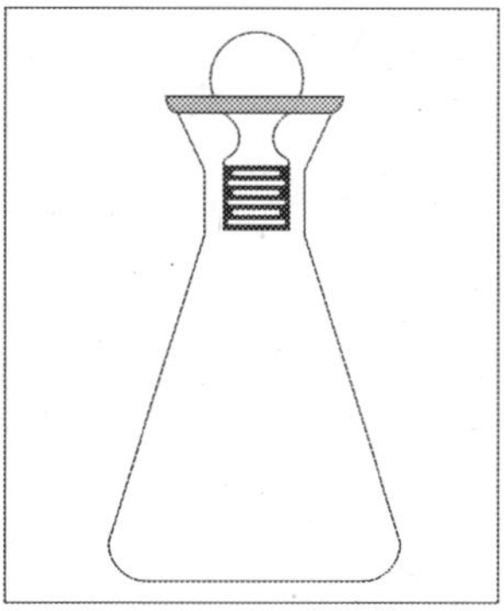

Iodine flask

Reagents:

1. Wiz's solution dissolve 8 g of ICl_3 in 200 ml of glacial acetic acid and 9 g of iodine in 400 ml of glacial acetic acid separately. Mix them together, dilute to 1000 ml with glacial acetic acid and store in amber colored bottles.
2. 10% KI solution
3. Starch indicator
4. 0.1 N Sodium thiosulfate

Sample/s Used: Oil samples

PROCEDURE

1. Weigh accurately by difference, an appropriate quantity of the oil and fat sample into a clean, dry 250 ml iodine flask (i.e., 2.5–3 g for coconut oil and 0.1–0.6 g for any other oil).

2. Add 10 ml of CCl_4, followed by addition of 25 ml of Wiz's solution.
3. Replace the stopper after moistening with 10% KI solution and keep the flask in dark for 30 minutes.
4. Then, add 15 ml of 10% KI solution and then 100 ml of distilled water.
5. Titrate against 0.1 N $Na_2S_2O_3$ using starch as indicator. End point is the disappearance of blue color. Repeat twice for each sample and also carry out a blank determination.

OBSERVATIONS AND CALCULATIONS

Standardization of Sodium Thiosulfate (0.1N)

Titration of $Na_2S_2O_3$ vs. $Kr_2Cr_2O_4$

S. No.	*Initial reading (ml)*	*Final reading (ml)*	*Volume used (ml)*

Volume of $K_2Cr_2O_7$ solution used for each titration =

$$\text{Normality of } Na_2S_2O_3 = \frac{\text{wt. of } K_2Cr_2O_7 \times 1000}{\text{mL of } Na_2S_2O_3 \text{ used} \times 49.037}$$

Formula used

$$\textbf{Iodine value} = \frac{(\text{Titre value B} - \text{Titre value S}) \times N \times 12.69}{W}$$

S = Sample used
B = Blank
N = Normality of thiosulfate sol.
W = Weight of the sample taken
12.69 = Equivalent weight of Iodine

RESULTS

Table showing the experimental iodine values of different samples.

S. No.	*Sample*	*Wt. of sample (g)*	*Titer value*			*Iodine value**
			Initial reading	*Final reading*	*Volume used*	

Table showing the experimental results of IV with standard reference value

S. No.	*Sample*	*Iodine value*		*Reference*#
		Experimental	*Standard*#	

#Refer E at page 159

Iodine value, also called iodine number is the measure of the degree of unsaturation of a fat or an oil. Saturated fats and oils do not take up iodine; therefore their iodine value is zero; but unsaturated fats and oils take up iodine as these compounds contain molecules with double or triple bonds, which are very reactive towards iodine. The more iodine is attached, the higher is the iodine value and the more reactive, less stable, and more susceptible to oxidation and rancidification is the fat or oil. If the iodine number is between 0-70, it will be a fat and if the value exceeds 70, then it is an oil.

PRECAUTIONS

1. Iodine flask should be properly cleaned.
2. End point should be carefully noted.
3. Care should be taken during preparation of Wiz's solution.
4. Weighing of sample should be done by weight differential method.
5. Iodine monochloride is caustic. So handle the reagent with gloves.
6. For better results, perform the experiments without any time gap during addition of reagents as the liberated iodine is susceptible to oxidation by light.

5.3.9 Determination of Peroxide Value

INTRODUCTION

Peroxides (ROOH) are primary reaction products formed in the initial stages of oxidation, and therefore give an indication of the progress of lipid oxidation. One of the most commonly used methods to determine peroxide value, utilizes the ability of peroxides to liberate iodine from potassium iodide. The lipid is dissolved in a suitable organic solvent and an excess of KI is added:

$$ROOH + KI_{excess} \longrightarrow ROH + KOH + I_2$$

Once the reaction has gone to completion, the amount of ROOH that has reacted can be determined by measuring the amount of iodine formed. This is done by titration with sodium thiosulfate and a starch indicator:

$$I_2 + \text{Starch} + 2Na_2S_2O_3 \text{ (blue)} \longrightarrow 2NaI + \text{Starch} + Na_2S_4O_6 \text{ (colorless)}$$

The amount of sodium thiosulfate required to titrate the reaction is related to the concentration of peroxides in the original sample. Thus, the peroxide value is usually determined volumetrically which has been largely developed by Lea. There are a number of problems with the use of peroxide value as an indicator of lipid oxidation. Firstly, peroxides are primary products that are broken down in the latter stages of lipid oxidation. Thus, a low value of P.V. may represent either the initial or final stages of oxidation. Secondly, the results of the procedure are highly sensitive to the conditions used to carry out the experiment, so the test must always be standardized. This technique is an example of a measurement of the increase in concentration of primary reaction products. This method gives rapid results. Fresh oils usually have P.V. well below 10 mEq/kg.

APPLICATION

Peroxide value measures a transient product of oxidation or advanced oxidation which can be distinguished by measuring over time. Some of the disadvantages of peroxide estimation in food stuff are a 5 g fat or oil sample size is required; which is difficult to obtain from food low in fat. The method is highly empirical; modification can change results. Despites drawbacks, P.V. is one of the most common tests of lipid oxidation.

REACTION INVOLVED

$R^{\cdot} + O_2 \longrightarrow RO_2$

$RO_2 + RH \longrightarrow ROOH + R^*$

$I^- + H_2O + ROOH \longrightarrow ROH + 2\,OH^- + I_2$

$I_2 + \text{Starch} + 2\,Na_2\,S_2\,O_3 \longrightarrow 2Na\,I + \text{Starch} + Na_2\,S_4\,O_6$

(Blue) (colorless)

FORMULA USED

$$P.V. = \frac{(\text{Titer value S} - \text{Titer value B}) \times N \times 1000}{W}$$

milliequivalents or
milllimoles per 1000 g of fat/oil
where

S = Sample used
B = Blank
N = Normality of thiosulfate sol.
W = Weight of the sample taken in g

MATERIALS REQUIRED

Glassware: Stoppered conical flask, Measuring cylinder, Pipette, Burette, Weighing balance

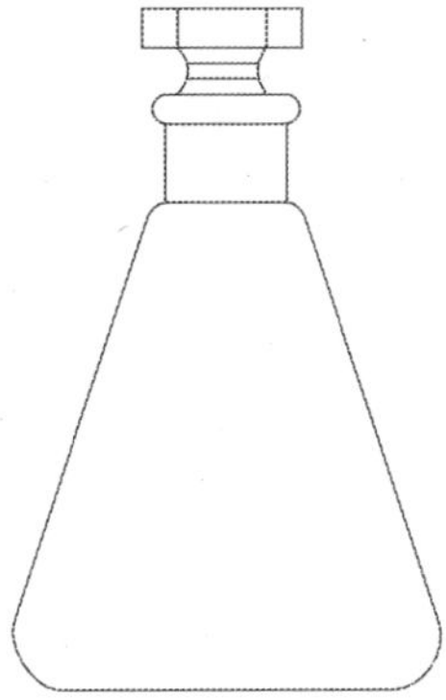

Reagents:

1. **Solvent:** Mix 3 parts by volume of glacial acetic acid with 2 parts by volume of chloroform.
2. **Saturated potassium iodide solution:** Dissolve 4 parts of pure potassium iodide in 3 parts of distilled water (freshly boiled and cooled). Store the solution in a brown bottle.

3. 0.1 N Sodium thiosulfate solution
4. 1% Starch indicator

Sample/s Used: Fat/oil samples (used and fresh both)

PROCEDURE

1. Weigh 5 g of fat sample in a stoppered conical flask.
2. To the sample, add 30 ml of solvent and dissolve.
3. To the flask add 0.5 ml of saturated KI and allow the solution to stand exactly for one minute with occasional shaking. Then, add 30 ml of distilled water followed by 0.5 ml of starch indicator.
4. Titrate against 0.1 N thiosulfate till a colorless end point. Repeat twice for a given sample and one blank minus sample.

Note: If the titer value is < 0.5 ml repeat the procedure using 0.01 N sodium thiosulfate soution.

OBSERVATIONS

Standardization of Sodium Thiosulfate using $Kr_2Cr_2O_4$

S. No.	*Initial reading*	*Final reading*	*Volume used*

Volume of $K_2Cr_2O_7$ solution used for each titration =

$$\text{Normality of } Na_2S_2O_3 = \frac{\text{Wt. of } K_2Cr_2O_7 \times 1000}{\text{mL of } Na_2S_2O_3 \text{ used} \times 49.037}$$

Table showing peroxide values of different experimental samples

S. No.	*Sample*	*Wt. of sample*		*Titer value (ml)*		*Peroxide Value**
		Fresh	*Used*	*Fresh*	*Used*	

Normality of thio. used =

CALCULATIONS

$$\text{*P.V.} = \frac{(\text{Titer value S} - \text{Titer value B}) \times N \times 1000}{W}$$

milliequivalents or
milllimoles per 1000 g of fat/oil

where

S = Sample used

B = Blank

N = Normality of thiosulfate sol.

W = Weight of the sample taken in g

RESULTS

S. No.	*Sample*	*Peroxide Value (millieq./1000 g)*

The peroxide value (P.V.) of an oil or fat is used as a measurement of the extent to which rancidity reactions have occurred during storage.

The molecular structure of fats and oils play an important role in autoxidation. Oils with a high degree of unsaturation are most susceptible to autoxidation. The best test for autoxidation (oxidative rancidity) is determination of the peroxide value, as peroxides are intermediates in the autoxidation reaction. Autoxidation is a reaction involving oxygen that leads to deterioration of fats and oils which form off-flavours and off-odours. Peroxide value, which is the concentration of peroxide in an oil or fat, is useful for assessing the extent to which spoilage has occurred.

The peroxide value is defined as the amount of peroxide oxygen per 1 kilogram of fat or oil. Traditionally this was expressed in units of milliequivalents, which has been commonly abbreviated as mequiv or even as meq. Correlation of rancid taste and peroxide value depends on the type of oil and is best tested with taste panels. The odors and flavors associated with typical oxidative rancidity are mostly due to carbonyl-type compounds.

It should be noted that peroxide values are not static and care must be taken in handling and when testing samples. It is difficult to provide a specific guideline relating peroxide value to rancidity. Peroxide values of fresh oils are less than 10 milliequivalents/kg. When the peroxide value is between 30 and 40 milliequivalents/kg, a rancid taste is noticeable. High peroxide values are a definite indication of a rancid fat, but moderate values may be the result of depletion of peroxides after reaching high concentrations.

PRECAUTIONS

1. Apparatus should be clean and absolutely dry.
2. Use stoppered conical flask to prevent escape of iodine vapours.
3. After adding KI, keep the flask aside for one minute exactly.
4. End point should be noted carefully.

5.3.10 Determination of Reichert-Meissl and Polenske Value

INTRODUCTION

Milk fat, like butter, can be distinguished from other fats by the presence of glycerol esters of relatively low molecular weight like butyric, capric, caprylic, lauric, etc., which are wholly or partially steam volatile.

The Reichert-Meissl (RM) value is the milliliters of 0.1 N aqueous alkali solution required to neutralise steam volatile, water soluble fatty acids distilled from 5 gm of oil/fat under the prescribed conditions. These are chiefly butyric and caproic acid.

The Polenske value (PV) is the number of milliliters of 0.1 N aqueous alkali solution required to neutralise steam-volatile, water-insoluble fatty acids distilled from 5 gm of oil/fat under the prescribed conditions. The test measures chiefly caprylic, capric, and lauric acid.

PRINCIPLE

The sample is saponified by heating with glycerol and sodium hydroxide. It is then split by dil. sulfuric acid. The volatile acids are immediately steam distilled which are filtered out and estimated by titration with standard sodium hydroxide solution.

MATERIALS REQUIRED

Apparatus: Heat source, Weighing balance, Whatman filter paper no. 4, Tubings

Glassware: Beaker, Graduated cylinder, Pipette, Burette, Bunsen burner, Glass beads, Condenser, Distillation flask, Thermocondenser, Conical flask, Round bottom flask with ground joints

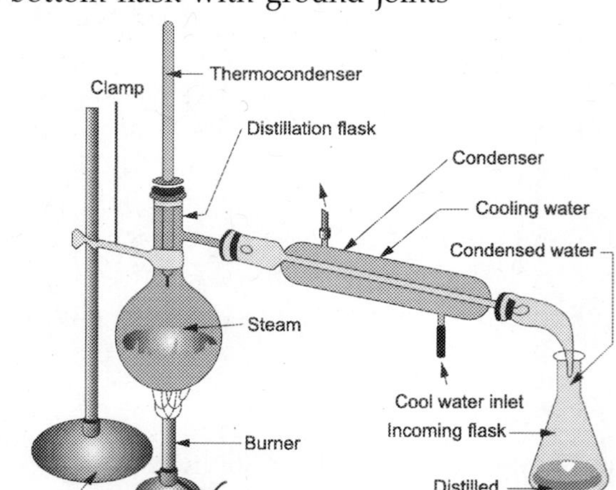

Experimental set-up for RM and PV determination

Reagents:

1. Glycerine
2. Conc NaOH
3. dil H_2SO_4
4. 0.1 N NaOH
5. Phenolphthalein
6. Ethyl alcohol

PROCEDURE

I. Determination of RM Value

1. Weigh 5 gm of filtered oil/fat sample in a clean and dry 300 ml distillation flask.
2. Add 20 ml of glycerine and 2 ml of concentrated NaOH solution to it. Heat it, swirling over a flame until completely saponified.
3. Cool the contents slightly and add 90 ml of boiling distilled water which has been vigorously boiled for 15 min to free it from CO_2. Mix thoroughly to make a clear solution.
4. If the solution is not clear or dark yellow then saponification needs to be repeated with fresh oil/fat sample. The dark coloration can occur if the sample is old.
5. To it, add a few glass beads and 50 ml dil. H_2SO_4. Connect the flask to distillation apparatus immediately to avoid the flame impinging on the surface of the flask above the liquid and cause super-heating.
6. Heat gently to liberate and separate the melted fatty acid.
7. Set the flame to collect 110 ml of distillate within 20 minutes. The distillate should be collected from the first drop as it falls from the condenser in the receiving graduated flask.
8. The water should be flowing at a sufficient speed to maintain the temperature of the water at outlet 15°C to 20°C.
9. Remove the flame immediately when the distillate reaches 110 ml mark of the flask. Quickly replace it by a 25 ml measuring cylinder.
10. Insert stopper in the graduated flask and place it in the water bath, maintained at 15°C, without mixing the contents of the flask. The flask should be placed such that its 110 ml mark is 1 cm submerged in water in the water bath.
11. Swirl the contents from time to time. Remove the flask from the water bath and dry its surface. Invert the flask 4 to 5 times to mix the contents but without shaking. Wetting of the stopper with insoluble acids should be avoided.

12. Filter the liquid through a dry Whatman filter paper no. 4.
13. Reject the first 2-3 ml of the filtrate and collect the rest in a dry flask. The filtrate should be clear.
14. Now, pipette 100 ml of the filtrate and add 5 ml of phenolphtahalein solution to it.
15. Titrate against standard 0.1 N NaOH solution.
16. Run a blank test, i.e., without fat sample using the same quantities of reagents.

OBSERVATIONS

S. No.	*Initial reading in the burette (in ml)*	*Final reading in the burette (in ml)*	*Titer value (in ml)*

CALCULATIONS

Reichert Meissl Value = (A-B) × N × 11

where,

A = volume of NaOH (in ml)

B = volume of NaOH required for the blank (in ml)

N = Normality of standard NaOH solution

II. Determination of Polenske Value

1. After titrating the soluble volatile acids, detach the still head.
2. Rinse the condenser with successive 15 ml portions of cold distilled water. Each washing should pass separately through the measuring cylinder, 110 ml graduated flask and the filter paper. Allow the washing of the filter paper to pass through.
3. Discard all the washings.
4. Place a funnel on a clean conical flask.
5. Dissolve the insoluble fatty acids by three similar washings of the condenser through the measuring cylinder, 110 ml stopper flask, and the filter paper with 15 ml portions of ethyl alcohol.
6. Combine all the alcoholic washings in a clean flask. Add 5 drops of phenolphtahalein indicator solution and titrate with standard 0.1 N NaOH.
7. Determine the Polenske value using the formula given below.

OBSERVATIONS

S. No.	*Initial reading in the burette (in ml)*	*Final reading in the burette (in ml)*	*Titer value (in ml)*

CALCULATIONS

$$\textbf{Polenske Value} = 10 \times V \times N$$

where,

V = Volume of NaOH (in ml)

N = Normality of standard NaOH

RESULTS

The given sample was found to have RM value______.

The given sample was found to have Polenske value______.

A high RM value determines the presence of milk fat due to high amounts of low molecular weight, water soluble fatty acids. Therefore, products like butter have high RM value since they are rich in butyric acid glycderides.

Similarly, coconut oil and palm kernel oil are found to be high in low molecular weight water insoluble fatty acids like capric, caprylic and lauric acid. Hence, these fats are found to have higher Polenske value.

PRECAUTIONS

1. The oil/fat sample should not be old.
2. The outlet temperature of water should be ranging from 15°C-20°C during distillation.
3. Vigorous mixing should not be done at any step.
4. Chemicals should be handled carefully.

REVIEW QUESTIONS

Saponification Value

1. What is saponification value?
2. What is the significance of saponification value?
3. What is the formula for saponification value?
4. Why 56.01 is multiplied in saponification value?
5. How is saponification value is related to molecular weight?
6. Is saponification value a back titration or a direct titration?

7. If the saponification value is 150, what will you infer?
8. Why is the normality of HCl taken to calculate saponification value?
9. Give the derivation of the formula for saponification value.
10. What is the S.V. for coconut oil?

Peroxide Value

1. What does peroxide value signify?
2. What type of titration is involved in peroxide value estimation?
3. Describe the roles of the chemicals in the peroxide value estimation?
4. What is the difference between the estimation of peroxide value and iodine value?
5. Why peroxide value is calculated in milliequivalents?
6. State the relationship between peroxide value and rancidity.
7. What are the different types of end products produced during autooxidation?
8. How will you determine the peroxide value of a food sample?
9. What does the high peroxide value indicate?

Free Fatty Acids

1. What is the acid value?
2. What does %free fatty acids signify?
3. Give the reaction for % free fatty acid estimation.
4. State the standard free fatty acids values for refined and non-refined oils as per FSSA, 2006.
5. What does multiplication factor 28.2 signify?
6. Give the relationship between the smoke point and free fatty acid.
7. Give the derivation of the % free fatty acid and acid value formula?
8. Why % free fatty acid of soya oil is 1.25% and vegetable oil is 0.25%?
9. Give the role of neutralized alcohol.
10. Justify the following
    (i) Ethyl alcohol is acidic or basic.
    (ii) Refining of oil whether increases or decreases % FFA.
11. Give the method for oil refining.
12. Why do food items made from ground nut oil spoil easily?

Iodine Value

1. What is iodine value?
2. Give the principle for iodine value.

3. In the iodine value determination, why is the blank volume higher than that of the sample?
4. What does a high versus a low iodine value tell you about the nature of the sample?
5. What does iodine value measure?
6. Write the standard values of iodine for butter and desi ghee.
7. Is iodine estimation a back titration?
8. What is the function of potassium iodide in iodine value?
9. What is the equivalent weight of iodine?
10. What is the role of ICL in Iodine value?
11. What is the iodine value for Dalda?
12. Give the composition of Wij's solution.
13. Why Wij's solution is kept so long at the magnetic stirrer?
14. Give the amount of the sample taken in iodine value and why?
15. Why is the iodine value of linseed oil so high and so low for coconut oil?
16. Identify the samples having iodine value 95, 140, 110 and 170 respectively.
17. Give reasons whether the iodine value will increase or decrease if an oil is prone to autoxiation.
18. What is the weight of the soya oil and desi ghee samples taken for determining Iodine value?

RM-PV Value

1. What is purpose of estimating RM value?
2. Which fatty acids are estimated by Polenske value?

R.I.

1. What is R.I.?
2. How does temperature affect the R.I. of a fat/oil sample?
3. Why the pycnometer should be fully filled with oil prior to estimation?

5.4 PROTEINS

Proteins are compounds found in all living cells, in animals and plants. They play a variety of important roles and are essential to maintain the structure and function of all life forms. The word 'protein' derives from the Greek word **protos**, meaning "primary" or "first". Proteins are vital for the growth and repair, and their functions are endless. Each and every property that characterizes a living organism is affected by proteins, whether it is a bacteria or a human body.

Proteins perform many functions which are essential for life. The building blocks of proteins are the twenty naturally occurring amino acids. These amino acids are liberated when proteins are hydrolyzed. Proteins are the polymers of amino acids.

$$\text{Proteins} \xrightarrow{\text{hydrolysis}} \text{Peptides} \xrightarrow{\text{hydrolysis}} \text{Amino acids}$$

All proteins contain the elements carbon, hydrogen, oxygen, nitrogen and sulfur. Some of these may also contain phosphorus, iodine, and traces of metals like ion, copper, zinc and manganese.

Proteins are polymers of amino acids. Proteins differ from each other according to the type, number and sequence of amino acids that make up the polypeptide backbone. As a result, they have different molecular structures, nutritional attributes and physicochemical properties. Proteins are important constituents of foods for a number of different reasons. They are a major source of *energy*, as well as containing essential amino-acids, such as lysine, tryptophan, methionine, leucine, isoleucine and valine, which are essential to human health, but which the body cannot synthesize. Proteins are also the major structural components of many natural foods, often determining their overall texture, e.g., tenderness of meat or fish products. Isolated proteins are often used in foods as ingredients because of their unique functional properties, i.e., their ability to provide desirable appearance, texture or stability. Typically, proteins are used as gelling agents, emulsifiers, foaming agents and thickeners. Many food proteins are enzymes which are capable of enhancing the rate of certain biochemical reactions. These reactions can have either a favorable or detrimental effect on the overall properties of foods. Food analysts are interested in knowing the total concentration, type, molecular structure and functional properties of the proteins in foods.

5.4.1 Determination of Proteins using Kjeldahl Method

INTRODUCTION

The Kjeldahl method was developed in 1883 by a brewer called Johann Kjeldahl. The method is used for the quantitative determination of crude protein in all types of foods. A food is digested with a strong acid so that it releases nitrogen which can be determined by a suitable titration technique. The amount of protein present is then calculated from the nitrogen concentration of the food. The same basic approach is still used today, although a number of improvements have been made to speed up the process and to obtain more accurate measurements. It is usually considered to be the standard method of determining protein concentration. Because the Kjeldahl method does not measure the protein content directly, a *conversion factor (F)* is needed to convert the measured nitrogen concentration to a protein concentration. A conversion factor of 6.25 (equivalent to 0.16 g nitrogen per gram of protein) is used for many applications, however, this is only an average value, and each protein has a different conversion factor depending on its amino-acid composition. The Kjeldahl method can conveniently be divided into three steps: digestion, neutralization and titration.

PRINCIPLE

Digestion

The food sample to be analyzed is weighed into a *digestion flask* and then digested by heating it in the presence of sulfuric acid (an oxidizing agent which digests the food), anhydrous sodium sulfate (to speed up the reaction by raising the boiling point) and a catalyst, such as copper, selenium, titanium or mercury (to speed up the reaction). Digestion converts any nitrogen in the food (other than that which is in the form of nitrates or nitrites) into ammonia and other organic matter to CO_2 and H_2O. Ammonia gas is not liberated in an acid solution because the ammonia is in the form of the ammonium ion (NH_4^+) which binds to the sulfate ion (SO_4^{2-}) and thus remains in solution:

$$\text{N (food)} \longrightarrow (NH_4)_2SO_4$$

Neutralization/Distillation

After the digestion has been completed, the digestion flask is connected to a *receiving flask* by a tube. The solution in the digestion flask is then made alkaline by addition of sodium hydroxide, which converts the ammonium sulfate into ammonia gas:

$$(NH_4)_2SO_4 + 2\ NaOH \longrightarrow 2NH_3 + 2H_2O + Na_2SO_4$$

Ammonia gas that is formed is liberated from the solution and moves out of the digestion flask and into the receiving flask which contains an excess of boric acid. The low pH of the solution in the receiving flask converts the ammonia gas into the ammonium ion, and simultaneously converts the boric acid to the borate ion:

$$NH_3 + H_3BO_3 \text{ (boric acid)} \longrightarrow NH_4^+ + H_2BO_3^- \text{ (borate ion)}$$

Titration

The nitrogen content is then estimated by titration of the ammonium borate formed with standard sulfuric or hydrochloric acid, using a suitable indicator to determine the end point of the reaction.

$$H_2BO_3^- + H^+ \longrightarrow H_3BO_3$$

The concentration of hydrogen ions (in moles) required to reach the end point is equivalent to the concentration of nitrogen that was in the original food. The following equation can be used to determine the nitrogen concentration of a sample that weighs *m* grams using a *x*M HCl acid solution for the titration:

$$\%\ N = \frac{x \text{ moles}}{1000 \text{ cm}^3} \times \frac{(v_5 - v_\delta)\text{ cm}^3}{\text{mg}} \times \frac{14\text{ g}}{\text{moles}} \times 100$$

where *vs* and *vb* are the titration volumes of the sample and blank, and 14 g is the molecular weight of nitrogen N. A blank sample is usually ran at the same time as the material being analyzed to take into account any residual nitrogen which may be in the reagents used to carry out the analysis. Once the nitrogen content has been determined, it is converted to a protein content using the appropriate conversion factor:

$$\%\text{Protein} = F \times \%\ N.$$

Determination of proteins using Kjeldahl method

- Degradation Step - Sample + $H_2SO_4 \rightarrow (NH_4)_2SO_4(aq) + CO_2(g) + SO_2(g) + H_2O(g)$
- Liberation of ammonia - $(NH_4)_2SO_4(aq) + 2NaOH \rightarrow Na_2SO_4(aq) + 2H_2O(l) + 2NH_3(g)$
- Capture of ammonia - $B(OH)_3 + H_2O + NH_3 \rightarrow NH_4^+ + B(OH)_4^-$
- Back-titration: $B(OH)_3 + H_2O + Na_2CO_3 \rightarrow NaHCO_3(aq) + NaB(OH)_4(aq) + CO_2(g) + H_2O$

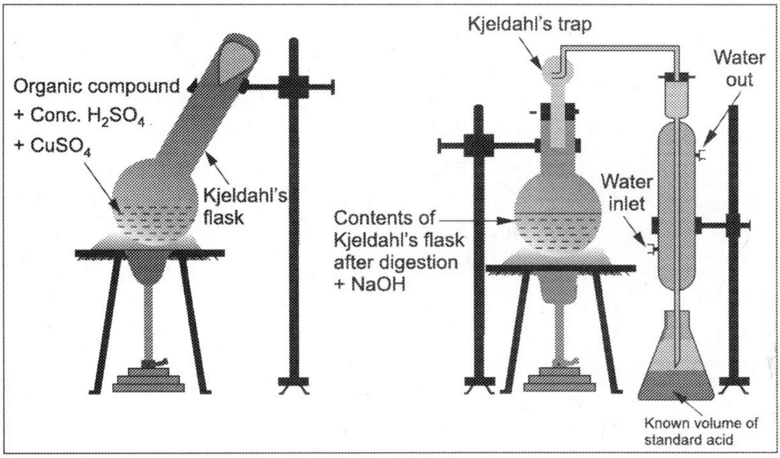

Kjeldahl's method

Advantages

The Kjeldahl method is widely used internationally and is still the standard method for comparison against all other methods. Its universality, high precision and good reproducibility have made it the major method for the estimation of protein in foods.

Disadvantages

It does not give a measure of the true protein, since all nitrogen in foods is not in the form of protein. Different proteins need different correction factors because they have different amino acid sequences. The use of concentrated sulfuric acid at high temperatures poses a considerable hazard, as does the use of some of the possible catalysts. The technique is time consuming to carry out.

MATERIALS REQUIRED

Apparatus: Kjeldahl's assembly, Erlenmeyer flask, Beaker, Pipettes, Burette stands, Glass beads

Reagents:

1. **Sulfuric acid, concentrated, A.R. grade or equivalent**
2. **Catalysts: Potassium sulfate and or copper sulfate with or without selenium dioxide**

   Note: For environmental and health concerns mercury containing catalysts should not be used.
3. **Sodium hydroxide solution, 50%. Weigh 500 g sodium hydroxide, dissolve in water and make up to 1 L with H_2O.**

   Note: The reaction of NaOH with water is exothermic. Place reaction vessel in a basin of ice cold water when dissolving it.

4. **Ammonium sulfate, A.R.**
5. **Hydrochloric acid, 0.1 N standard solution.** Pipette 8.3 ml of concentrated hydrochloric acid to approximately 500 ml distilled H_2O in 1 L volumetric flask soaked in ice cold water. Allow to cool and make up to volume with distilled H_2O. Pour the solution in 1 L brown bottle and let the solution stand for 2 or 3 days with occasional shaking before standardization.

   Standardization of HCl solution. Weigh accurately about 0.47 g sodium tetraborate decahydrate (borax) into a 250 ml conical flask, dissolve in about 50 ml water and add a few drops of methyl red indicator solution. Titrate with the hydrochloric acid from a burette until the color changes to pink at the end point.

$$\text{Normality} = \frac{\text{Wt. of borax} \times 100 \text{ ml}}{\text{Volume of HCl titrant} \times 190.72}$$

6. **Methyl red/bromcresol green indicator solution.** Dissolve 0.2 g methyl red and dilute to 100 ml with 95% ethanol. Dissolve 1.0 g bromcresol green and dilute to 500 ml with 95% ethanol. Mix one (1) part methyl red solution with five (5) parts bromcresol green solution.

   Note: Methylene blue, 0.2% in 95% ethanol and methyl red, 0.2% in ethanol (1:2) indicator can also be used.
7. **Boric acid solution, 4% with indicator.** Dissolve 40 g boric acid in hot distilled water, cool and make up to 1 L with distilled water and add 3 ml methyl red/bromcresol green indicator solution. Solution will be light orange in color.

PROCEDURE

1. **Preparation of sample:** Grind or blend the sample until homogeneous. If the sample cannot be analyzed on the same day, keep in screw-cap bottle in a freezer. For samples intended for analysis of vitamins or other labile nutrients, flush sample with nitrogen before storing.
2. **Analysis**
   (a) Blank: Include two reagent blanks (containing all reagents used in nitrogen analysis except the sample) in every batch of analysis to subtract reagent nitrogen from the sample nitrogen.

   (b) **Test sample**

   (i) Thaw out sample to room temperature and mix the sample thoroughly.

(ii) Weigh in duplicate 2–10 g sample (depending on the nitrogen content of the sample) into the digestion tube.

(iii) Add 5-7 g catalyst and 1 glass bead to prevent solution from bumping and 10–20 ml sulfuric acid.

(iv) Place digestion tube in the digestor. Digest mixture initially at low temperature to prevent frothing and boil briskly until the solution is clear and is free of carbon or until oxidation is complete.

Note: The digestion time and volume of sulfuric acid required depends on the material to be digested. If the digest is still yellowish, cool the digest and add an additional 5-10 ml sulfuric acid.

(v) Continue digestion until a clear digest is obtained.

(vi) Heat for another hour after the liquid has become clear to complete breakdown of all organic matter.

(vii) Place a 250 ml Erlenmeyer flask containing 50 ml of 4% boric acid with indicator as receiver on the distillation unit.

Note: The tip of the condenser should extend below the surface of the acid solution.

(viii) Add 100 ml of water and 70 ml of 50% sodium hydroxide to the digests and start distillation.

Note: Make sure excess NaOH is added to neutralize sulfuric acid to ensure complete release of ammonia.

(ix) Distill until all ammonia has been released or approximately >or equal to 150 ml distillate is obtained.

Note: Use condenser with ice cold water to effectively capture all distilled ammonia.

(x) Lower the receiver flask so that the delivery tube is above the liquid surface and continue the distillation for 1-2 minutes. Finally, rinse the delivery tube with water and allow the washings to drain into the flask.

(xi) Titrate the distillate with the standardised 0.1 N hydrochloric acid until the first appearance of the pink color.

(xii) Record volume of acid used to the nearest 0.05 ml.

Note: Always check the normality of the standardised 0.1 N HCl.

FORMULA USED

$$\%\ \text{Nitrogen} = \frac{(\text{ml standard acid} - \text{ml blank}) \times \text{N of acid} \times 1.4007}{\text{Weight of sample in grams}}$$

If the sample weight is in milligrams, the molecular weight of nitrogen should be changed to 1400.67.

If it is desired to determine % protein instead of % nitrogen, the calculated % N is multiplied by a factor, the magnitude of the factor depending on the sample matrix. Many protein factors have been developed for use with various types of samples as shown in the table below.

PROTEIN FACTOR

A factor is the used percent N to the percent crude protein. Most proteins contain 16 percent N so the conversion factor is 6.25 (100/16 = 6.25)

$$\frac{\%\,N}{0.16} = \%\text{ protein}$$

$$\%\,N \times 6.25 = \%\text{ protein}$$

Table showing factors for the conversion of nitrogen to protein

Food items	*Factor*
Meat and Fish	6.25
Gelatin	5.55
Milk and milk products	6.38
Casein	6.4
Human milk	6.37
Eggs: Whole	6.25
Vitellin	6.32
Albumin	6.12
Wheat: Whole	5.83
Bran	6.31
Embryo	5.8
Endosperm	5.7
Rice and rice flour	5.95
Barley, oats, rye and flour	5.83
Millet	6.31
Corn/Maize, beans	6.25
Soybean	5.71
Castor bean	5.30
Peanuts, Brazil nuts, groundnuts	5.46
Coconuts, cashew nuts and other nuts	5.30
Sesame seeds, sunflower seeds and all other seeds	5.30
All other Foods	6.25

OBSERVATION AND CALCULATIONS

Table showing % Nitrogen and % Protein

	Sample weight (g)	*Vol. HCL (ml) titrant*	*% Nitrogen**	*% Protein#*
Blank				
Sample				

FORMULA USED:

$$\% \text{ Nitrogen} = \frac{(\text{ml standard acid} - \text{ml blank}) \times \text{N of acid} \times 1.4007}{\text{Weight of sample in grams}}$$

% Protein = % Total nitrogen × appropriate nitrogen conversion factor

Note: Report test results (in g per 100 g sample) to one decimal place.

RESULTS

The % nitrogen of the given sample was found to be ______ and % protein to be ______.

PRECAUTIONS

1. Always wear protective devices, e.g., face mask, goggles and gloves when handling toxic and corrosive chemicals.
2. Understand the toxicity and safety of the reagents used before starting the method.
3. Use pipette aides in handling acids.
4. Digest sample under a hood to reduce inhalation of acid fumes.
5. Analysis should be done in an ammonia-free environment.

5.4.2 Determination of Proteins using Biuret Method

INTRODUCTION

A violet-purplish color is produced when cupric ions are complexed with peptide bonds (CO-NH). Substances containing at least two peptide bonds, for example, biuret, large peptides, and all proteins react under alkaline conditions to produce this color. The absorbance of the color produced is read at 540 nm. The color intensity (absorbance) is proportional to the protein content of the sample.

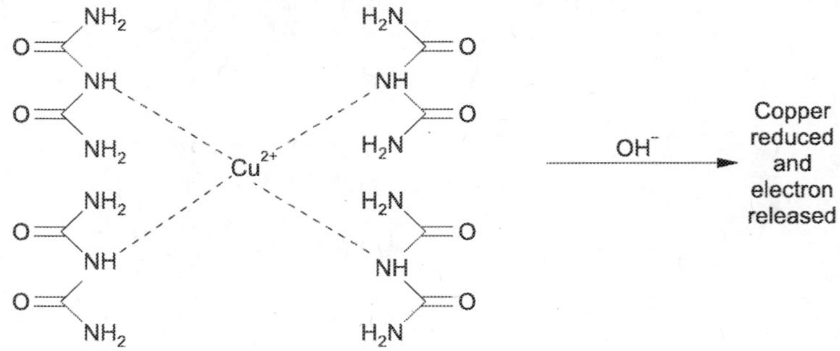

The biuret method has been used to determine proteins in cereal, meat, soybean proteins, and as a qualitative test for animal feed. The biuret method also can be used to measure the protein content of isolated proteins.

ADVANTAGES

1. Less expensive than the Kjeldahl method; rapid (can be completed in less than 30 min); simplest method for analysis of proteins.
2. Color deviations are encountered less frequently than with Lowry Ultraviolet (UV) absorption or turbid metric methods.
3. Very few substances other than proteins in foods interfere with the biuret reaction.
4. It does not detect nitrogen from nonpeptide or nonprotein sources.
5. It is not sensitive to pH change.

DISADVANTAGES

1. Not very sensitive as compared to the Lowry method; requires at least 2-4 rng protein for assay.
2. Absorbance could be contributed from bile pigments if present.
3. High concentration of ammonium salts interfere with the reaction.
4. Color varies with different proteins; gelatin gives a pinkish-purple color.

5. Opalescence could occur in the final solution if high levels of lipid or carbohydrate are present.
6. Not an absolute method: color must be standardized against known protein (e.g., BSA) or against the Kjeldahl nitrogen method.

MATERIALS REQUIRED

Apparatus: Spectrophotometer, Centrifuge, Burette stand

Glassware: Beaker, Pipette, Conical flask

Reagents:

1. **Biuret reagent ($CuSO_2.5H_2O$, Na-K tartarate + KI)**

   The biuret reagent is made of sodium hydroxide (NaOH) and hydrated copper(II) sulfate, together with potassium sodium tartrate. Potassium sodium tartrate is added to complex to stabilize the cupric ions.
2. **Standard Protein**

   Dissolve 50 mg BSA in 50 ml of distilled water in a volumetric flask. Take 10 ml of this stock standard and dilute to 50 ml in another flask for working standard solution. One ml of this solution contains 200 μg protein.

Sample/s Used: Protein sample

PROCEDURE

1. A 3 ml biuret reagent is mixed with a 1 ml portion of protein solution (1-10 mg protein/ml).

   The reagent includes copper sulfate, NaOH, and potassium sodium tartrate, which is used to stabilize the cupric ion in the alkaline solution.
2. After the reaction mixture is allowed to stand at room temperature for 15-30 min, the absorbance is read at 540 nm against a reagent blank.
3. Filtration or centrifugation before reading absorbance is required if the reaction mixture is not clear.
4. A standard curve of concentration versus absorbance is constructed using bovine serum albumin (BSA) as shown below.

S. No.	*Test tube*	*Vol. of protein sol^n (5 mg/ml)*	*Vol. of H_2O*	*Vol. of biuret reagent*	*Absorbance at 540 nm*
1.	B	–	2 ml	3 ml	
2.	S-1	0.4 ml	1.6 ml	3 ml	
3.	S-3	1.2 ml	0.8 ml	3 ml	
4.	S-5	2 ml	–	3 ml	

Sample B					
1.	T-1	1 ml	1 ml	3 ml	
2.	T-2	1 ml	1 ml	3 ml	

S = BSA, T = Test sample, B = Black

RESULTS

The amount of protein in the unknown sample is _______mg/ml.

PRECAUTIONS

1. Allow the reaction mixture to stand for 15 minutes at least.
2. Filter or centrifuge the contents of the reaction mixture if not clear prior to taking absorbance.

5.4.3 Determination of Proteins using Lowry's Method

INTRODUCTION

The phenolic group of tyrosine and trytophan residues (amino acid) in a protein will produce a blue purple color complex, with maximum absorption in the region of 660 nm wavelength, with Folin- Ciocalteau (FC) reagent which consists of sodium tungstate molybdate and phosphate. Thus, the intensity of color depends on the amount of these aromatic amino acids present and will thus, vary for different proteins. Most protein estimation techniques use Bovine Serum Albumin (BSA) universally as a standard protein because of its low cost, high purity and ready availability. The method is sensitive down to about 10 μg/ml and is probably the most widely used protein assay despite its being only a relative method, subject to interference from Tris buffer, EDTA, nonionic and cationic detergents, carbohydrate, lipids and some salts. The incubation time is very critical for a reproducible assay. The reaction is also dependent on pH and a working range of pH 9 to 10.5 is essential.

The principle behind the Lowry's method of determining protein concentrations lies in the reactivity of the peptide nitrogen[s] with the copper [II] ions under alkaline conditions and the subsequent reduction of the Folin-Ciocalteau reagent phosphomolybdic phosphotungstic acid to heteropolymolybdenum blue by the copper-catalyzed oxidation of aromatic acids.

It is a two-step reaction:

Ist step: Oxidation of aromatic amino acid

$$\text{Protein} + CuSO_4^- \xrightarrow{OH^-} \text{Protein complex}$$

2nd step: Reduction of FC reagent

$$\text{Protein complex} \xrightarrow[\text{FC reagent}]{\text{reduction}} \text{Blue color (Molybdenum tungsten blue complex)}$$

Advantages

1. It is a sensitive assay which requires no digestion.
2. It is more sensitive to pH change. Therefore, the pH of assay is maintained at 10-15 pH.
3. It is sensitive to low concentration of protein (0.1-2 mg)
4. 10-20 times more sensitive than measurement of the UV at 280 nm.

Disadvantages

1. Narrow pH range within which it is accurate.
2. The procedure is sensitive because it employs two color forming reactions. Blue color is formed by reduction of FC by amino acids.

Then, the color intensity depends on the amount of these aromatic amino acids present.

MATERIALS REQUIRED

Apparatus: Colorimeter

Glassware: Test tubes, Pipettes, Conical flask

Reagents:

1. **Reagent A:** 2% sodium carbonate in 0.1 N sodium hydroxide.
2. **Reagent B:** 0.5% copper sulfate ($CuSO_4.5H_2O$) in 1% potassium sodium tartarate. Prepare fresh by mixing stock solutions.
3. **Alkaline copper solution (Reagent C):** Mix 50 ml of reagent A and 1 ml of reagent B prior to use.
4. **Diluted Folin's reagent (Reagent D):** Dilute Folin-Ciocalteau reagent with an equal volume of 0.1 N NaOH
5. **Standard:** Dissolve 50 mg BSA in 50 ml of distilled water in a volumetric flask. Take 10 ml of this stock standard and dilute to 50 ml in another flask for working standard solution. One ml of this solution contains 200 μg protein.

PROCEDURE

1. Different dilutions of BSA solutions are prepared by mixing stock BSA solution (1 mg/ml) and water in the test tube as given in the table. The final volume in each of the test tubes is 4 ml. The BSA range is 0.05 to 1 mg/ml.
2. Then add 5.5 ml of alkaline copper sulfate solution (Reagent C). Mix the solutions well.
3. This solution is incubated at room temperature for 10 mins.
4. Then add 0.5 ml of Folin Ciocalteau solution (Reagent D) to each tube and incubate for 30 min. Adjust the colorimeter with blank to zero and take the optical density at 660 nm.
5. Plot the absorbance against protein concentration to get a standard calibration curve.
6. Check the absorbance of unknown sample and determine the concentration of the unknown sample using the standard curve.

Table showing prepation of standard curve for determining concentration unknown protein sample by Lowry's method

BSA (mg/ml)	*Distilled water (ml)*	*Reagent C (leave for 10 min.)*	*Folin's reagent (leave for 30 min.)*	*O.D. at 660 nm*
0.1	3.9	5.5	0.5	
0.2	3.8	5.5	0.5	
0.3	3.7	5.5	0.5	
0.4	3.6	5.5	0.5	
0.5	3.5	5.5	0.5	
0.6	3.4	5.5	0.5	
0.7	3.3	5.5	0.5	
0.8	3.2	5.5	0.5	
0.9	3.1	5.5	0.5	
1.0	3.0	5.5	0.5	
Unknown Sample (ml)				
0.1	3.9	5.5	0.5	
0.2	3.8	5.5	0.5	

RESULTS

The amount of protein in the unknown sample is ________ mg/ml.

PRECAUTIONS

1. The protein samples should be stored under refrigerated conditions.
2. The cuvettes should be cleaned and dried properly after every reading.

REVIEW QUESTIONS

1. Could phenolphthalein be used as an indicator in the Kjeldahl titration?
2. Describe the function of the following chemicals used in Kjeldahl determination:
   (i) Catalyst pellet (ii) Borate
   (iii) H_2SO_4 (iv) NaOH
3. Why is it not necessary to standardize the boric acid solution?
4. Explain how the factor used to calculate the percent protein for a corn product was obtained, and why the protein factors for some other cereal grains (e.g., wheat, oats) differ from that of corn.
5. What are the three steps involved in the Kjeldahl method of protein analysis?

6. Why milliliters of HCl can be used as an indirect measure of the protein content of a sample?
7. Why is the conversion factor from Kjeldahl nitrogen to protein different for various foods, and how is the factor of 6.25 obtained?
8. Which test is used for qualitative estimation of proteins in meat?
9. What is the advantage of biuret test?
10. What is the composition of Folin-Ciocalteau reagent?

5.5 CARBOHYDRATES

Carbohydrates are one of the most important components in many foods. Carbohydrates may be present as isolated molecules or they may be physically associated or chemically bound to other molecules. Individual molecules can be classified as follows according to the number of monomers that they contain.

Monosaccharides

Monosaccharides are water-soluble crystalline compounds. They are aliphatic aldehydes or ketones which contain one carbonyl group and one or more hydroxyl groups. Most natural monosachharides have either five (pentoses) or six (hexoses) carbon atoms. Commonly occurring hexoses in foods are glucose, fructose and galactose, whilst commonly occurring pentoses are arabinose and xylose. The reactive centers of monosaccharides are the carbonyl and hydroxyl groups.

Oligosaccharides

These are relatively low molecular weight polymers of monosaccharides (< 20) that are covalently bonded through glycosidic linkages. Disaccharides consist of two monomers, whereas trisaccharides consist of three. Oligosaccharides containing glucose, fructose and galactose monomers are the most commonly occurring in foods.

Polysaccharides

The majority of carbohydrates found in nature are present as polysaccharides. Polysaccharides are high molecular weight polymers of monosaccharides (> 20). Polysaccharides containing all the same monosaccharides are called homopolysaccharides (e.g., starch, cellulose and glycogen are formed from only glucose), whereas those which contain more than one type of monomer are known as heteropolysaccharides (e.g., pectin, hemicellulose and gums).

Carbohydrates also contribute to the sweetness, appearance and textural characteristics of many foods apart from being an important source of energy and dietary fiber.

Carbohydrate analysis of raw materials and processed foods can be used to provide a wealth of important information, e.g., the so-called fingerprint oligosaccharide patterns can be used to detect any adulteration. In addition, sugar breakdown products can be used to determine if a food has been irradiated. Therefore, it is important to determine the type and concentration of carbohydrates in foods for a number of reasons.

Need for testing of carbohydrates

(i) **Standards of Identity:** Foods must have compositions which conform to government regulations.

(ii) **Nutritional Labeling:** To inform consumers of the nutritional content of foods.

(iii) **Detection of Adulteration:** Each food type has a carbohydrate "fingerprint".

(iv) **Food Quality:** Physicochemical properties of foods such as sweetness, appearance, stability and texture depend on the type and concentration of carbohydrates present.

(v) **Economic:** Industry doesn't want to give away expensive ingredients.

(vi) **Food Processing:** The efficiency of many food processing operations depends on the type and concentration of carbohydrates that are present.

5.5.1 Determination of Total Sugars using Lane Eyon Method

INTRODUCTION

Total sugar in this method refers to all sugars that undergo sugar inversion with hydrochloric acid. The method involves the inversion of sugars present in food samples with hydrochloric acid. The sugar present in a specified volume of the hydrolysed solution is used to reduce copper in the Fehling's solution previously standardised with working standard invert sugar solution. Excess copper is back titrated with the standard sugar solution. The difference in the volume of standard sugar used for the standardization and for back titration is a measure of total sugar content of the sample.

MATERIALS REQUIRED

Apparatus: Burette, Bunsen burner, Desiccator with dessicant, Conical flask

Reagents:

1. **Fehling stock solution A:** Dissolve 69.3 g Copper (II) Sulfate pentahydrate ($CuSO_4$, $5H_2O$) in distilled water to 1000 ml.
2. **Fehling stock solution B:** Dissolve 346 g potassium sodium tartrate tetrahydrate ($KNaC_4H_4O_6$ $4H_2O$) and 100 g sodium hydroxide (NaOH) in distilled water to 1000 ml. Decant solution from sediment if necessary before use.
3. **Methylene blue indicator, 1% aqueous solution.**
4. **Anhydrous D-glucose:** Dry for two hours at 100°C and cool in dessicator before use.
5. **Standard glucose solution:** Weigh 1.25 g dried anhydrous D-glucose and dissolve in distilled water to 250 ml volume.

Sample/s Used: Different types of sugars

PROCEDURE

I. Standardization

1. Make mixed Fehling's solution by transferring first 12.5 ml of solution A and then 12.5 ml of solution B to a 250 ml conical flask and mix well.
2. Fill 50 ml burette with standard glucose solution. Add from burette about 23.5 ml of the standard glucose solution to mixed Fehling's solution. Add a few glass beads and place on wire gauze. Heat over Bunsen burner. Boil for two minutes and add three drops of methylene blue indicator.

3. Then, without interrupting the boiling, add 2-3 drops of glucose solution at about 10 seconds intervals until the blue color completely disappears.
4. Titration should be completed in less than a minute. Repeat titration twice.

   Note: Exactly 24.1 ml of standard glucose solution should be used for the titration, otherwise adjust Fehling's stock solution A appropriately and repeat standardization.

II. Determination

1. Weigh accurately into a beaker (g) g sample, where approximate sample weight (g) = 12500/(expected DE × dry substance).
2. Dilute sample with distilled water to 250 ml.
3. Proceed as described above.
4. Standardization using sample solution instead of standard glucose solution for the titration.

   Note: The test solution should contain approximately 1.25 g dextrose in 250 ml.

CALCULATIONS

$$\text{DE (dextrose equvalent)} = \frac{(25 \times a \times 10)}{b \times g \times d};$$

where

a = factor from table below

b = ml test solution for titration

g = gram sample dissolved to 250 ml

d = % dry solids in sample

Note: DE (average results of two samples to one decimal)

S. No.	*ml of sugar solution required*	*Dextrose factor ~mg dextrose corresponding to 25 ml of Fehling's solution*	*mg dextrose per 100 ml*
1.	15	120.2	801
2.	16	120.2	751
3.	17	120.2	707
4.	18	120.2	668
5.	19	120.3	633
6.	20	120.3	601.5
7.	21	120.3	572.9
8.	22	120.4	547.3
9.	23	120.4	523.6

10.	24	120.5	501.9
11.	25	120.5	482.0
12.	26	120.6	463.7
13.	27	120.6	446.8
14.	28	120.7	431.1
15.	29	120.7	416.4
16.	30	120.8	402.7
17.	31	120.8	389.7
18	32	120.8	377.6
19.	33	120.9	366.3
20.	34	120.9	355.6
21.	35	121.0	345.6
22.	36	121.0	336.3
23.	37	121.1	327.4
24.	38	121.2	318.8
25.	39	121.2	310.7
26.	40	121.2	303.1
27.	41	121.3	295.9
28.	42	121.4	289.0
29.	43	121.4	282.4
30.	44	121.5	276.1
31.	45	121.5	270.1
32.	46	121.6	264.3
33.	47	121.6	258.8
34.	48	121.7	253.5
35.	49	121.7	248.4
36.	50	121.8	243.6

RESULTS

The mg dextrose per 100 ml of the sample is__________.

PRECAUTIONS

1. Understand the toxicity and safety of the reagents used before starting the method.
2. Use safety devices such as tong or gloves in handling hot containers.

5.5.2 Determination of Total Carbohydrates by Anthrone Method

INTRODUCTION

Carbohydrates are the important components of storage and structural materials in the plants.They exist as free sugars and polysaccharides. The basic units of carbohydrates are the monosaccharides which cannot be split by hydrolysis into more simpler sugars. The carbohydrate content can be measured by hydrolysing the polysaccharides into simple sugars by acid hydrolysis and estimating the resultant monosaccharides.

PRINCIPLE

Carbohydrates are first hydrolysed into simple sugars using dilute hydrochloric acid. In hot acidic medium, glucose is dehydrated to hydroxymethyl furfural. This compound forms with anthrone, a green colored product with an absorption maximum at 630 nm.

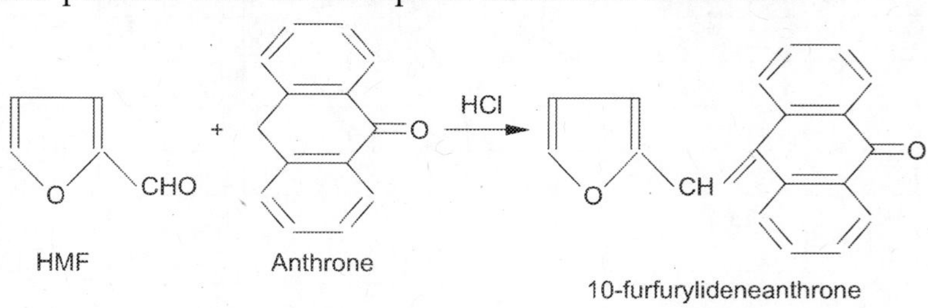

MATERIALS REQUIRED

Apparatus: Spectrophotometer, Centrifuge, Water bath

Glassware:

(i) Boiling tubes

A boiling tube is a small cylindrical vessel used to strongly heat substances in the flame of a Bunsen burner. A boiling tube is essentially a scaled-up test tube, being about 50% larger. They are designed to be wide enough to allow substances to boil violently as opposed to a test tube, which is too narrow; a boiling liquid can explode out of the end of test tubes when they are heated, as there is no room for bubbles of gas to escape independently of the surrounding liquid. Boiling tubes are made from borosilicate glass that can withstand high temperatures.

(ii) Conical flask

Reagents:

1. **2.5 N HCl**
2. **Anthrone reagent:** Dissolve 200 mg anthrone in 100 ml of ice-cold 95% H_2SO_4. Prepare fresh before use.

3. **Standard glucose standard:** Dissolve 100 mg in 100 ml water.
4. **Working standard:** 10 ml of standard diluted to 100 ml with distilled water. (Store refrigerated after adding a few drops of toluene.)

Sample/s Used: Different sugar solutions

PROCEDURE

1. Weigh 100 mg of the sample into a boiling tube.
2. Hydrolyse by keeping it in a boiling water bath for three hours with 5 ml of 2.5 N HCl and cool to room temperature.
3. Neutralise it with solid sodium carbonate until the effervescence ceases.
4. Make up the volume to 100 ml and centrifuge.
5. Collect the supernatant and take 0.5 and 1 ml aliquots for analysis.
6. Prepare the standards by taking 0, 0.2, 0.4, 0.6, 0.8 and 1 ml of the working standard. '0' serves as blank.
7. Make up the volume to 1 ml in all the tubes including the sample tubes by adding distilled water.
8. Then add 4 ml of anthrone reagent.
9. Heat for eight minutes in a boiling water bath.
10. Cool rapidly and observe the green to dark green color at 630 nm.
11. Draw a standard graph by plotting concentration of the standard on the X-axis versus absorbance on the Y-axis.
12. From the graph, calculate the amount of carbohydrate present in the sample tube.

CALCULATIONS

Amount of carbohydrate present in 100 mg of the sample

$$= \frac{\text{mg of glucose}}{\text{volume of test sample}} \times 100$$

RESULTS

The amount of carbohydrate present in 100 mg of the sample is______.

PRECAUTION

1. Cool the contents of all the tubes on ice before adding ice-cold anthrone reagent.
2. Heat the contents for eight minutes then cool rapidly before taking O.D.

5.5.3 Estimation of Total Carbohydrates by Phenol Sulfuric Acid Method

INTRODUCTION

Phenol sulfuric acid method is the easiest and most reliable method amongst the quantitative assays for carbohydrate estimation. It is mostly used for measuring neutral sugar content in oligosaccharides, proteoglycans, glycoproteins and glycolipids. In hot acidic medium, glucose is dehydrated to hydroxymethyl furfural. This forms a yellow-brown colored product with phenol and has absorption maximum at 490 nm. This is one of the best methods to estimate total carbohydrate.

MATERIALS REQUIRED

Apparatus: Spectrophotometer, Water bath

Glassware: Test tube, Pipettes, Conical flask

Reagents:

1. **Phenol 5%:** Redistilled (reagent grade) phenol (50 g) dissolved in water and diluted to one liter.
2. Sulfuric acid 96%, reagent grade.
3. **Standard glucose:** Standard—100 mg of glucose in 100 ml of water. *Note: Different sugars can also be used.*
4. **Working standard:** 10 ml of standard diluted to 100 ml with distilled water.

Sample/s Used: Sugar solutions

PROCEDURE

1. Follow the steps 1 to 4 as given in anthrone method for sample preparation.
2. Pipette out 0.2, 0.4, 0.6, 0.8 and 1 ml of the working standard into a series of test tubes.
3. Pipette out 0.1 and 0.2 ml of the sample solution in two separate test tubes. Make up the volume in each tube to 1 ml with water.
4. Set a blank with 1 ml of water.
5. Add 1 ml of phenol solution to each tube.
6. Add 5 ml of 96% sulfuric acid to each tube and shake well.
7. After 10 minutes shake the contents in the tubes and place in a water bath at 25–30°C for 20 min.
8. Read the color at 490 nm (for pentoses, 480 nm will be used).
9. Calculate the amount of total carbohydrate present in the sample solution using the standard graph.

CALCULATIONS

Absorbance corresponds to 0.1mL of the test = 'X' mg of glucose

100 mL of the sample solution contains = $\frac{\text{'X'}}{0.1} \times 100$ mg of glucose

contains = % of total carbohydrate present

RESULTS

The % total carbohydrate present in the given sample is________.

PRECAUTIONS

1. For hexoses, the O.D. must be taken at 490 nm and for pentoses at 480 nm.
2. Water bath temperature should be between 25 and 30°C.

5.5.4 Determination of Reducing Sugars by Nelson-Somogyi Method

INTRODUCTION

Sugars with reducing property (arising out of the presence of a potential aldehyde or keto group) are called reducing sugars. Some of the reducing sugars are glucose, galactose, lactose and maltose. The Nelson-Somogyi method is one of the classical and widely used methods for the quantitative determination of reducing sugars.

PRINCIPLE

The reducing sugars when heated with alkaline copper tartrate reduce the copper from the cupric to cuprous state and thus cuprous oxide is formed. When cuprous oxide is treated with arsenomolybdic acid, the reduction of molybdic acid to molybdenum blue takes place. The blue color developed is compared with a set of standards in a colorimeter at 620 nm.

MATERIALS REQUIRED

Apparatus: Water bath, Vortex mixer, Colorimeter, Weighing balance

Glassware: Pipettes, Conical flask

Reagents:

1. **Alkaline Copper Tartrate**
   - (i) Solution A: Dissolve 2.5 g anhydrous sodium carbonate, 2 g sodium bicarbonate, 2.5 g potassium sodium tartrate and 20 g anhydrous sodium sulfate in 80 ml water and make up to 100 ml.
   - (ii) Solution B: Dissolve 15 g copper sulfate in a small volume of distilled water. Add one drop of sulfuric acid and make up to 100 ml.

   Note: Mix 4 ml of B and 96 ml of solution A before use.
2. **Arsenomolybdate reagent:** Dissolve 2.5 g ammonium molybdate in 45 ml water. Add 2.5 ml sulfuric acid and mix well. Then add 0.3 g disodium hydrogen arsenate dissolved in 25 ml water. Mix well and incubate at 37°C for 24–48 hours.
3. **Standard glucose solution:** 100 mg glucose in 100 ml distilled water.
4. **Working standard:** 10 ml of standard diluted to 100 ml with distilled water [100 µg/ml].
5. **Ethanol (80%)**

Sample/s Used: Sugar solution

PROCEDURE

1. Weigh 100 mg of the sample and extract the sugars with warm 80% ethanol, twice (5 ml each time).
2. Collect the supernatant and evaporate it by keeping it on a water bath at 80°C.
3. Add 10 ml water and dissolve the sugars.
4. Pipette out aliquots of 0.1 or 0.2 ml to separate test tubes.
5. Pipette out 0.2, 0.4, 0.6, 0.8 and 1 ml of the working standard solution into a series of test tubes.
6. Make up the volume in both sample and standard tubes to 2 ml with distilled water.
7. Pipette out 2 ml distilled water in a separate tube to set a blank.
8. Add 1 ml of alkaline copper tartrate reagent to each tube.
9. Place the tubes in a boiling water for 10 minutes.
10. Cool the tubes and add 1 ml of arsenomolybolic acid reagent to all the tubes.
11. Make up the volume in each tube to 10 ml with water.
12. Read the absorbance of blue color at 620 nm after 10 min.
13. From the graph drawn, calculate the amount of reducing sugars present in the sample.

CALCULATIONS

Absorbance corresponds to 0.1 ml of test = X mg of glucose

10 ml contains = $0.1X \times 10$ mg of glucose = % of reducing sugars.

X value is obtained from standard graph

RESULTS

The % reducing sugars is found to be ___________.

PRECAUTIONS

1. Extraction of the sugars should be done carefully with warm ethanol.
2. Take absorbance after 10 minutes of mixing of reagents so that the blue color develops.

5.5.5 Estimation of Reducing Sugars by Dinitrosalicylic Acid (DNSA) Method

PRINCIPLE

3, 5-Dinitrosalicylic acid (DNSA) is used extensively in biochemistry for the estimation of reducing sugars. It detects the presence of free carbonyl group (C=O) of reducing sugars. This involves the oxidation of the aldehyde functional group (in glucose) and the ketone functional group (in fructose). During this reaction, DNSA is reduced to 3-Amino- 5-Nitrosalicylic acid (ANSA) which under alkaline conditions is converted to a reddish brown colored complex which has an absorbance maximum of 540 nm.

$$\text{Aldehyde group} \xrightarrow{\text{oxidation}} \text{Carboxyl group}$$

$$\text{3, 5-Dinitrosalicylic acid} \xrightarrow{\text{reduction}} \text{3-Amino, 5-Nitrosalicylic acid}$$

MATERIALS REQUIRED

Apparatus: Water bath, Colorimeter/Spectrophotometer, Weighing balance

Glassware: Test tubes, Conical flask, Burette

Reagents:

1. **Dinitrosalicylic Acid Reagent (DNS Reagent)**
   (i) Dissolve by stirring 1 g dinitrosalicylic acid, 200 mg crystalline phenol and 50 mg sodium sulphite in 100 ml 1% NaOH. Store at 4°C. Since the reagent deteriorates due to sodium sulphite, if long storage is required, sodium sulphite may be added at the time of use.
2. **40% Rochelle salt solution (potassium sodium tartrate).**

Sample/s Used: Different sugars

PROCEDURE

1. Follow, steps 1 to 3 as in Nelson-Somogyi's method to extract the reducing sugars from the test material.
2. Pipette out 0.1 to 1 ml of the extract in test tubes and equalize the volume to 1 ml with water in all the tubes.
3. Add 3 ml of DNS reagent.
4. Heat the contents in a boiling water bath for 5 min.
5. When the contents of the tubes are still warm, add 1 ml of 40% Rochelle salt solution.

6. Cool and read the intensity of dark red color at 510 nm.
7. Run a series of standards using glucose as shown in the table below and plot a graph.

	Glucose solution	*Sample*	*Water*	*DNS reagent cover the tubes (with aluminum foil) and heat for 5 min. in a boiling water bath*	*Soduim potasuim tartarate*
B	–	–	1	3	1
1	0.1	–	0.9	3	1
2	0.2	–	0.8	3	1
3	0.3	–	0.7	3	1
4	0.4	–	0.6	3	1
5	0.5	–	0.5	3	1
6	0.6	–	0.4	3	1
7	0.7	–	0.3	3	1
8	0.8	–	0.2	3	1
9	0.9	–	0.1	3	1
10	1	–	–	3	1
S1	–	0.4	0.6	3	1
S2	–	0.6	0.4	3	1

OBSERVATIONS

Table showing absorbance and its relationship with CHO content

Tube	*Absorbance*	*CHO content (mg/dl)*
B	–	–
1		2
2		4
3		6
4		8
5		10
6		12
7		14
8		16
9		18
10		20
S1		
S2		

CALCULATIONS

Calculate the amount of reducing sugars present in the sample using the standard graph.

- Dilution factor = $\dfrac{\text{Final volume}}{\text{Aliquot volume}}$
- D.F Sample 1 = $\dfrac{1}{0.4}$ = 2.5
- D.F Sample 2 = $\dfrac{1}{0.6}$ = 1.7

RESULTS

The amount of carbohydrate in 1 gram of sample = _____ mg/dl × dilution factor × 100

PRECAUTIONS

1. Add the Rochelle salt solution when the contents are still warm.
2. The dilution factor should be accounted for in the calculation.

5.5.6 Estimation of Reducing and Non-Reducing Sugars using Potassium Ferricyanide Method

Sugars in a basic solution (pH > 10.5) can reduce ferricyanide to ferrocyanide. The amount is equivalent to the amount of sugar present in the sample. The remaining ferricyanide reacts with KI and the untreated KI is volumetrically measured with standard thiosulfate solution using starch as an indicator. This method is called alkaline potassium ferricyanide method and is commonly used for cereals and for their products.

MATERIALS REQUIRED

Apparatus: Water bath, Whattman no. 4.

Glassware: Burette, Pipette, Conical flask, Stoppered test tubes, Volumetric flask

Reagents:

1. **Acetate Buffer (pH 4.4 - 4.6):** Dissolve 4.1 g of anhydrous sodium acetate in little water. To this add 3 ml of glacial acetic acid and make up the volume to 1 L.
2. **Alkaline Ferricyanide solution:** Dissolve 22 g of Na_2CO_3 in little amount of water. To this, add 16.5 g of pure dry K_4 [Fe $(CN)_4$] and make up the volume to 500 ml.
3. **Acetic Acid Salt Solution:** Add 70 g of KCl in distilled water and completely dissolve, then add 40 g of $ZnSO_4$ to it. (In case, KCl is not dissolving, keep the volumetric flask in a lukewarm water bath.). To this add 200 ml of glacial acetic acid and make up the volume to 1 L.
4. **50% of KI solution:** Dissolve 50 g of KI in 100 ml of distilled water.
5. **10% H_2SO_4 solution:** Take 10 ml of conc. H_2SO_4 and make up the volume to 100 ml.
6. **12% Sodium tungstate:** Take 12 g of sodium tungstate and dissolve in 100 ml of distilled water.
7. **0.1 N Sodium thiosulfate**
8. **1% Starch indicator:** Dissolve 1 g of starch in lukewarm 100 ml distilled water, boil and then cool the contents.

Sample/s Used: Maida, Wheat flour, Soya flour

Table showing role of different reagents used

Reagent used	*Role*
Acetate buffer	To extract sugar, enzymes and proteins.
Conc. sulfuric acid	To inhibit the enzymes.
Sodium tungstate	To denature the proteins.
Potassium ferricyanide	To oxidize self and reduce the sugars.
Acetic acid salt solution	Helps in precipitation of potassium ferricyanide as it interacts with the zinc ions to form a brick red zinc complex.

Reactions Involved

$$2K_3\,Fe\,(CN)_6 + 2KOH \rightleftharpoons 2K_4\,Fe\,(CN)_6 + H_2O$$

$$CH_2OH\,(CHOH)_4\,CHO + 1/2\,O_2 \rightleftharpoons \underset{\text{Gluconic acid}}{CH_2OH\,(CHOH)_4\,COOH}$$

$$K_3\,Fe\,(CN)_6 + {}^{*}KI \rightleftharpoons K_4\,[Fe\,(CN)_6] + I$$

$$3\,ZnSO_4 + 2K_4\,Fe\,(CN)_6 \rightleftharpoons K_2\,Zn_3\,[Fe\,(CN)_6]_2 + 3K_2SO_4$$

PROCEDURE

1. Take 5 g of the given sample in 250 ml conical flask.
2. To this add 46 ml of acetate buffer followed by 2 ml of 10% H_2SO_4 and 2 ml of 12% sodium tungstate solution.
3. Mix the contents thoroughly and filter using Whattman paper no. 4.
4. Collect the filtrate in two stoppered test tubes each and the test tubes are labeled as A and C.
5. In the similar manner, without addition of sample, collect the filtrate in a stoppered test tube and label as B (blank).
6. TUBE A: To 5 ml of filtrate, add 10 ml of $K_4\,[Fe(CN)_4]$. Incubate in boiling water bath for 20 min. Cool the tube in running water and then transfer the contents to a conical flask containing 25 ml of acetic acid salt solution. Then, add 1 ml of 50% KI and 1 ml of starch indicator. Titrate the contents against standard thiosulfate (.1N) solution to a white colored end point.
7. TUBE B: Follow the same procedure as that of A.
8. TUBE C: Incubate 5 ml of filtrate in boiling water bath for 15 min. Cool the contents under running tap water and proceed as A.

FORMULA USED

(a) For reducing sugars: mg of maltose/10 g of sample = $(B–A)$

(b) For non-reducing sugar: mg of sucrose/10 g of sample = $(B–C) – (B–A) = A–C$

where

A = Tube A titer value

B = Tube B titer value

C = Tube C titer value

OBSERVATIONS

S. No.	*Sample*	*Sodium thio. titer values (ml)*		
		A	*B*	*C*

CALCULATIONS

(a) Reducing sugars: mg of maltose/10 g of sample = $(B–A)$

(b) Non-reducing sugar: mg of sucrose/10 g of sample = $(B–C) – (B–A) = A–C$.

where

A = Tube A titer value

B = Tube B titer value

C = Tube C titer value

(c) Calculate the reducing and non-reducing content using the following table.

Ferricyanide-Maltose/Sucrose Conversion Table

0.1 N Ferricyanide reduced (ml)	*Maltose/ 10 g Flour (mg)*	*Sucrose/ 10 g Flour (mg)*	*0.1 N Ferricyanide reduced (ml)*	*Maltose/10 g Flour (mg)*	*Sucrose/ 10 g Flour (mg)*
0.10	5	5	4.60	244	218
0.20	10	10	4.70	251	223
0.30	15	15	4.80	257	228
0.40	20	19	4.90	264	233
0.50	25	24	5.00	270	238
0.60	31	29	5.10	276	242
0.70	36	34	5.20	282	247
0.80	41	38	5.30	288	251
0.90	46	43	5.40	295	256
1.00	51	48	5.50	302	261

Contd...

Contd...

1.10	56	52	5.60	308	266
1.20	60	56	5.70	315	270
1.30	65	62	5.80	322	275
1.40	71	67	5.90	328	280
1.50	76	71	6.00	334	285
1.60	80	76	6.10	341	290
1.70	85	81	6.20	347	294
1.80	90	86	6.30	353	299
1.90	96	91	6.40	360	304
2.00	101	95	6.50	367	309
2.10	106	100	6.60	373	313
2.20	111	104	6.70	379	318
2.30	116	109	6.80	385	323
2.40	121	114	6.90	392	328
2.50	126	119	7.00	398	333
2.60	130	123	7.10	406	337
2.70	135	128	7.20	412	342
2.80	140	133	7.30	418	347
2.90	145	138	7.40	425	352
3.00	151	143	7.50	431	357
3.10	156	148	7.60	438	362
3.20	161	152	7.70	445	367
3.30	166	157	7.80	451	372
3.40	171	161	7.90	458	377
3.50	176	166	8.00	465	382
3.60	182	171	8.10	472	387
3.70	188	176	8.20	478	392
3.80	195	181	8.30	485	397
3.90	201	185	8.40	492	402
4.00	207	190	8.50	499	407
4.10	213	195	8.60	505	–
4.20	218	200	8.70	512	–
4.30	225	204	8.80	519	–
4.40	231	209	8.90	–	–
4.50	237	214			

RESULTS

The reducing and the non-reducing sugar contents in the given samples are as follows:

S. No.	*Sample*	*Reducing sugar (mg of maltose/10 g of sample)**	*Non-reducing sugar (mg of maltose/10 g of sample)**
1.			
2.			

* : The values are taken from the Ferricyanide Maltose/Sucrose conversion table

PRECAUTIONS

1. Glassware should be properly cleaned.
2. Incubation time should be carefully noted.
3. $ZnSO_4$ should be added after complete solubilization of KCl.
4. Cooling should be done quickly.

5.5.7 Estimation of Crude Fiber

Crude fiber consists largely of cellulose and lignin plus some mineral matter. It represents around 60–80% of the cellulose and 4–6% of the lignin. The crude fiber content is commonly used as a measure of the nutritive value of poultry and livestock feeds and also in the analysis of various foods and food products to detect adulteration, quality and quantity.

PRINCIPLE

During the acid and subsequent alkali treatment, oxidative hydrolytic degradation of the native cellulose and considerable degradation of lignin occur. The residue obtained after final filtration is weighed, incinerated, cooled and weighed again. The loss in weight gives the crude fiber content.

MATERIALS REQUIRED

Apparatus: Water bath, Muffle Furnance Dessicator

Glassware: Silica crucible, Conical flask, Pipettes

Reagents:

1. **Sulfuric acid solution (0.25 N):** 1.25 g concentrated sulfuric acid diluted to 100 ml (concentration must be checked by titration).
2. **Sodium hydroxide solution (0.3 N):** 1.25 g sodium hydroxide in 100 ml distilled water (concentration must be checked by titration with standard acid).
3. **Petroleum ether:** AR grade

Sample/s Used: Plant Foods

PROCEDURE

1. Extract 2 g of ground material with ether or petroleum ether to remove fat initial boiling temperature 35–38°C and final temperature 52°C). If fat content is below 1%, extraction may be omitted.
2. After extraction with ether, boil 2 g of dried material with 200 ml of sulfuric acid for 30 min with glass beads.
3. Filter through muslin cloth and wash with boiling water until washings are no longer acidic.
4. Boil with 200 ml of sodium hydroxide solution for 30 min.
5. Filter through muslin cloth again and wash with 25 ml of boiling 1.25% H_2SO_4, three 50 ml portions of water and 25 ml alcohol.
6. Remove the residue and transfer to silica crucible (preweighed dish W_1).

7. Dry the residue for 2 h at 130 ± 2°C. Cool the dish in a desiccator and weigh (W_2).
8. Ignite for 30 min at 600 ± 15°C.
9. Cool in a desiccator and reweigh (W_3).

CALCULATIONS

% Crude fiber in ground sample

$$= \frac{\text{Loss in weight on igntion } [(W_2 - W_1) - (W_3 - W_1)]}{\text{Weight of the sample}} \times 100$$

RESULTS

The % crude fiber in the given sample is_________.

PRECAUTIONS

1. Extraction of the sample is must but it can be omitted if the fat content is less than 1%.
2. Washing should be repeated four to five times.

REVIEW QUESTIONS

1. What are reducing and non-reducing sugars and give their examples?
2. Give the principle of reducing sugar and non-reducing sugar estimation.
3. Give the structural difference between reducing and non-reducing sugar.
4. Name the carbon atom at which the reducing sugar group is present.
5. What is the position of free aldehydic and ketonic group in glucose and fructose?
6. Give the role of potassium ferricyanide in sugar estimation.
7. What method is used for determining reducing sugars in fruit juices?
8. What is the principle behind determination of total reducing sugars using the Somogyi–Nelson method?
9. The Munson–Walker, Lane–Eynon, and Somogyi–Nelson methods can be used to measure reducing sugars. Give the similarities and differences among these methods.
10. Distinguish chemically between monosaccharides, oligosaccharides, and polysaccharides.

11. Classify each as a reducing or non-reducing carbohydrate: D-glucose, D-fructose, sorbitol, sucrose, maltose, raffinose, cellulose, amylopectin.
12. What are the principles behind total carbohydrate determination using the phenol-sulfuric acid method?

5.6 PIGMENTS

The quality of food is generally based on color, flavor, aroma, texture and nutritive values. However, no matter how nutritious and tasty the food is, it is unlikely to be eaten unless it has the "right" color. The colors of food are contributed by natural pigments, or by artificial colorants/ color additives.

Pigment is a term generally referring to normal constituents of cells or tissues that impart color. Chlorophylls, anthocyanins, flavonoids, proanthocyanidins, tannins, betalains, quinones, xanthones, carotenoids, myoglobin and hemoglobin are the common natural food pigments occurring in plants and animals.

A few categories of pigments are listed below along with their characteristic range of colors.

Pigment type	*Colors*
Anthocyanins (subclass of flavonoids)	Blue/Purple/Red
Anthoxanthins (subclass of flavonoids)	Yellow - Ivory
Betacyanins	Yellow - Red/Purple
Carotenoids	Yellow - Red
Chlorophylls	Greens
Xanthophylls (a subclass of carotenoids)	Ivory - Yellow

The pigments are widely present in plant sources both fruits and vegitables as shown:

Pigment	*Found in*
Anthocyanins	Berries, Grapes, Red Peppers, Beets, Eggplant, Plums
Beta-Carotene	Carrots, Pumpkin, Sweet Potatoes, Citrus, Papaya, Melon, Squash
Curcumin	Turmeric
Lutein	Broccoli, Spinach
Lycopene	Tomatoes, Watermelon, Red Grapefruits
Zeaxanthin	Corn

In nature, attractive coloration of flowers and fruits by anthocyanin pigments help attract pollinators and hence facilitates plant pollinations. Besides this function, anthocyanins in photosynthetic tissues can protect cells from strong light intensity damage by absorbing blue-green light, thereby protecting the tissues from photoinhibition. Apart from these, anthocyanins are also powerful antioxidants as they have a strong reducing property. This antioxidant property remains even after the anthocyanins in plant are consumed by another organism. There are some supporting evidence shown in recent researches that anthocyanins in berries have potential health effects against cancer, aging, neurological diseases, inflammation, diabetes and bacterial infections.

5.6.1 Thin Layer Chromatography Separation of the Green Pigment Extracts from Spinach Leaves

INTRODUCTION

Chlorophyll is a green pigment found in most plants and algae. Chlorophyll absorbs light most strongly in the blue and red but poorly in the green portions of the electromagnetic spectrum, hence we can perceive the green color of chlorophyll-containing tissues like plant leaves.

In plant cells, these chlorophylls are located within plastid bodies called chloroplast. Chlorophylls are vital for photosynthesis, which allows plants to obtain energy from light. Chlorophyll is a chlorin pigment, which is structurally similar to other porphyrin pigments such as heme. At the center of the chlorin ring, is a magnesium ion. The chlorin ring can have several different side chains, usually including a long phytol chain.

There are several forms of chlorophyll existing in nature, such as chlorophylls *a, b, c* and *d*, bacteriochlorophylls *a* and *b*. In foods, we are mainly concerned with chlorophyll *a* and *b* which occur in the approximate ratio of 3:1 in terrestrial higher plants. Chlorophylls *a* and *b* are soluble in alcohols, ether, benzene and acetone, but are insoluble in water. Chemically, the chlorophylls may be altered in many ways, but in food processing, the most common alternation is pheophytinization, which is the replacement of the central magnesium by hydrogen and the consequent formation of the dull olive-brown phenophytins.

Thin layer chromatography is a fast, simple and generic chromatography technique applied to separate chemical compounds from a mixture. It involves a stationary phase consisting of a thin layer of adsorbent material, usually silica gel or cellulose, immobilized onto a flat and inert carrier plate.

A sample mixture, which consists of the different chemical compounds to be separated, is applied on the stationary phase and then dissolved in an appropriate solvent (mobile phase). The mixture will be carried upward along the plate due to the capillary action and the components will be separated based on the different polarities of the individual compounds.

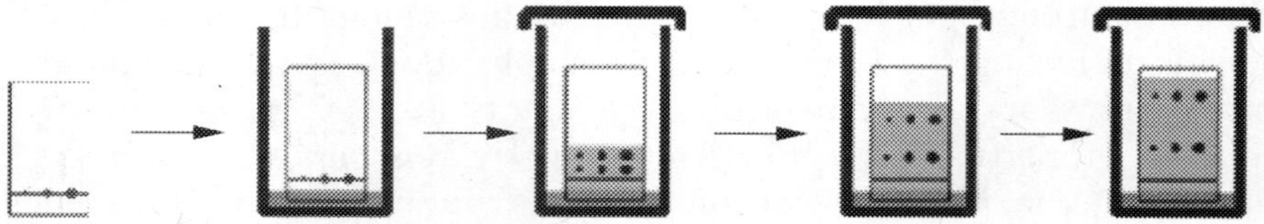

TLC plate development sequence

MATERIALS REQUIRED

Apparatus: Beakers, Glass-Rod, Funnel and Filter Paper, Conical Flask, Pre-coated Silica Gel TLC Plates, Glass Capillary, TLC Developing Tank, Hair Dryer, UV-Light

Reagents:

1. 85% Aqueous Acetone
2. Calcium Carbonate
3. TLC Developing Solvent (Petroleum Ether-Isopropanol-Water, 100:10:0.25 v/v/v)

Sample/s Used: Green Spinach Leaves, Coriander

PROCEDURE

1. Cut spinach leaves into fine pieces. Weigh an approximately 15 g portion.
2. Place them in a 100 ml beaker and add a small quantity (0.1 g) of calcium carbonate. Grind the tissue with a glass rod for a short time.
3. Add 30 ml 85% aqueous acetone to the tissue and continue to grind until a fine pulp mixture is formed.
4. Transfer the pulp mixture to a funnel with filter paper and filter. Collect the pigment extract.
5. With the aid of a glass capillary tube, spot a small aliquot of the pigment extract on a silica gel TLC plate. The spot should be approximate 1.5 cm from the end of the TLC plate. Keep the sample spot as small as possible (no more than 5 mm in diameter).
6. If necessary, dry the spot with a hair dryer during spotting. Re-apply aliquots of the pigment extract if the color of the spot is too pale, until the spot is dark enough.
7. Repeat steps 1-6 with another vegetable leaf sample (coriander). Spot the pigment extract on the same TLC plate. It should be noted that the two spots should be located on the same horizontal level and at least 1-1.5 cm apart from each other.
8. When the sample spots are thoroughly dry, place the TLC plate in a developing chamber containing the running solvent (petroleum ether-isopropanol-water, 100:10:0.25 v/v/v) at a level just below the sample spots. The height of the solvent is crucial as the sample spots must not drip into the solvent.
9. Cover the chamber with a lid. Let the solvent run upwards and do not disturb the chamber.
10. When the solvent front has approached near the top end of the TLC plate, use a forceps to remove the TLC plate from the developing chamber.

11. Allow the TLC plate to air dry (hair dryer may be used but warm air should be applied gently).
12. View the TLC plate both visually by eye and under UV light. Record the relative position and color of all the spots on the TLC plate.
13. Compare the pigment spectrum of the samples in the different lanes.

RESULTS

1. Draw arrows to the locations of the solvent front and the colored bands. Label each band as to its pigment. (Beta-carotene is the most non-polar pigment (highest R_f) and its band will be yellow. Chlorophyll *a* has a larger R_f than does chlorophyll *b*.
2. For the following calculations mark the center of the initial pigment dot; this will be the starting point for all the following measurements. Also mark the middle point of each pigment band and the solvent front.
3. The rate at which a pigment moves up the plate is reported as an R_f value which is defined as the ratio of the distance moved by the spot to the distance moved by the solvent. Determine the R_f values for each of the pigments you observe using the formula provided below.

R_f value

The retention factor, or R_f, is defined as the distance traveled by the compound divided by the distance traveled by the solvent.

$$R_f = \frac{\text{Distance traveled by the compound}}{\text{Distance traveled by the solvent front}}$$

For example, if a compound travels 2.1 cm and the solvent front travels 2.8 cm, the R_f is 0.75:

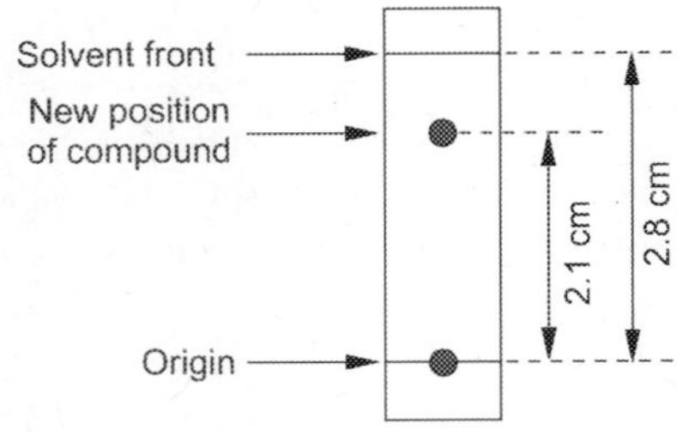

$$R_f = \frac{2.1}{2.8} = 0.75$$

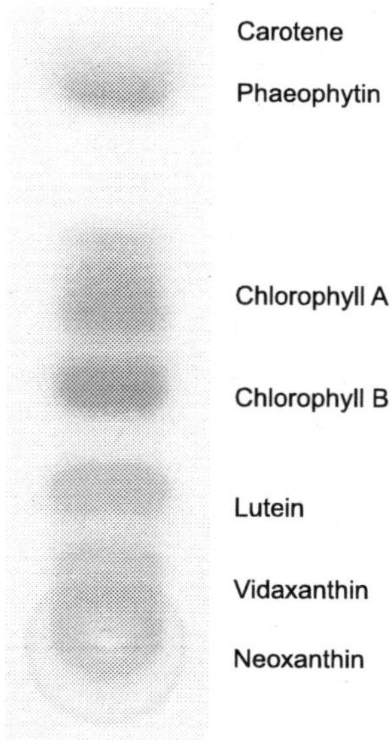

TLC separation of spinach pigments

PRECAUTIONS

1. The TLC solvent is volatile and should be handled carefully.
2. The used solvent should be collected in a waste bottle.

5.6.2 Effect of pH on Food Anthocyanins

INTRODUCTION

Anthocyanins are another group of pigments commonly found in edible plants. They are a group of water-soluble vacuolar flavonoid pigments that are widespread in fruits, vegetables and flowers. They are synthesized by organisms of the plant kingdom and have been observed to occur in all tissues of higher plants, providing attractive color to leaves, stems, roots, flowers, and fruits.

All anthocyanins are derivatives of the basic flavylium cation structure. More than 550 anthocyanins are known but only a few (e.g., Pelargonidin, Cyanidin, Delphinidin, Peonidin, Petunidin and Malvidin) are important in food as others are comparatively rare and are found in non-edible flowers and leaves. The flavylium nucleus of the anthocyanin pigment is electron-deficient and therefore highly reactive. The reactions usually involve decolorization of the pigments and are usually undesirable in fruit and vegetable processing.

Anthocyanins are commonly found in vegetables and fruits in reddish color, such as blackcurrant, raspberry, cherry, eggplant, blue grape, red/black beans and red cabbage, and the amounts are relatively large. One kilogram of blackberry contains approximately 1.15 g and red/black legumes can contain up to 20 g/kg.

One of the interesting properties of anthocyanins is that they show a marked color change with different pH, being red in acids and blue in bases. It involves different chemical structural changes in response to the pH variation.

MATERIALS REQUIRED

Apparatus: Test tubes, Micropippete with tips, Vortex mixer, Timer

Reagents: pH Buffer of 2.0, 3.0, 5.0, 7.0, 9.0 and 10.0

Sample/s Used: White grape juice and Black grape juice

PROCEDURE

1. Place 10 ml portions of buffer solutions of the following pH values: 2.0, 3.0, 5.0, 7.0, 9.0 and 10.0 in separate test tubes.
2. To each buffer solution, add 1 ml of the black grape juice and mix well. Keep it for 3 min and then observe any color change.
3. Repeat the test by using a white grape juice sample. Record the results.

OBSERVATIONS

	Color of juice samples at buffer					
	pH 2.0	*pH 3.0*	*pH 5.0*	*pH 7.0*	*pH 9.0*	*pH 10.0*
Grape Juice (at 0 min)						
Grape Juice (at 10 min)						
White Grape Juice (at 0 min)						
White Grape Juice (at 10 min)						

RESULTS

Increasing pH causes greater destruction of anthocyanin in samples. Flavylium salts are stable only in highly acidic conditions. These salts lose the proton in higher pH and transform into quinoidal base, which is an unstable pigment, and immediately bond to water and form colorless compound called chromenol.

PRECAUTIONS

1. Buffer solutions with different pH should be prepared accurately.
2. The samples should be completely immersed in the buffer solution.

REVIEW QUESTIONS

1. Name some water-soluble food pigments.
2. Based on their absorption spectra, would you expect chlorophylls *a* and *b* to be the same color? Explain.
3. Why should you not use ink on the coating to mark your pigment placement?
4. In the TLC experiment, why must the spot be applied to the TLC plate above the level of the developing solvent?
5. What will be the result of applying too much compound to a TLC plate?
6. What is the effect of increasing pH on anthocyanins?

5.7 VITAMINS

Vitamins are organic compounds or set of organic compounds that are found in natural substances found in plants and animals. They are known as essential nutrients for human beings because of their requirement. It was originally thought that these organic substances were only amines and so the name is given by shortening **"vital amines" to "vitamins."**

Vitamins can be divided into fat-soluble and water soluble groups, vitamins *A, E, D* and K are fat-soluble and may be stored in the body, the remainder being water-soluble and the body has limited or no stores.

Vitamins are classified according to their biological and chemical activity and not according to their structure. There are 13 essential vitamins known so far and each one has a special role to play in helping to regulate the processes such as cell growth and repair, reproduction and digestion and support sexual function in human beings.

The thirteen vitamins are listed below:

1. **Vitamin A:** It is a fat-soluble vitamin and it contains retinol, retinal, and four carotenoids including beta carotene.
2. **Vitamin B_1:** This is a water-soluble vitamin and contain a compound called thiamine having amine group.
3. **Vitamin B_2:** This is a water-soluble vitamin containing a compound called riboflavin.
4. **Vitamin B_3:** A water-soluble vitamin containing compound niacin or niacinamide, a nitrogen containing vitamin.
5. **Vitamin B_5:** Contains pantothenic acid and is a water-soluble vitamin.
6. **Vitamin B_6:** This vitamin is a combination of compounds pyridoxine, pyridoxamine and pyridoxal. This is also a water-soluble vitamin.
7. **Vitamin B_7:** This vitamin contains a compound called biotine. This is also a water-soluble vitamin.
8. **Vitamin B_9:** Folic acid and folinic acid are the components of this vitamin which is water-soluble.
9. **Vitamin B_{12}:** This vitamin is a combination of cyanacobalamin, hydroxycobalamin and methyl cobalamin. This is a water-soluble vitamin.

   All these vitamins belonging to Vitamin B class are in a combination known as **B COMPLEX**. All these water-soluble vitamins are absorbed directly by the body and excess is excreted.
10. **Vitamin C:** A water-soluble vitamin containing ascorbic aid.
11. **Vitamin D:** This a fat-soluble vitamin containing cholecalciferol.

12. **Vitamin E:** This vitamin is a combination of chemicals belonging to tocoferols and tocotrienols. This is also a fat soluble vitamin.
13. **Vitamin K:** Philloquinones and menaquinones are the compounds in this vitamin and this is a fat soluble vitamin.

All the fat-soluble vitamins are stored in the fat of the body and recovered from fat when they are needed. Hence, overdose of these vitamins should be avoided.

Table showing sources of vitamins

Vitamin	*Sources*
Vitamin A	Leafy Vegetables, Carrots, Pumpkin, Orange, Amla, Ripe Yellow Fruits, Spinach and Liver.
Vitamin B_1	Brown Rice, Potatoes, Vegetables, Oat Meal, Liver, Eggs and Pork
Vitamin B_2	Asparagus, Banana, Popcorn, Green Beans and Milk Products.
Vitamin B_3	Mushrooms, Many Vegetables, Dry Nuts, Meat, Fish and Eggs
Vitamin B_5	Avocados, Meat, Broccoli
Vitamin B_6	Dry Nuts, Bananas, Green Vegetables and Meat
Vitamin B_7	Peanuts, Vegetables, Raw Egg Yolk and Liver
Vitamin B_9	Bread, Cereal, Pasta, Leafy Vegetables and Liver
Vitamin B_{12}	Animal Source of Food Items such as Meat
Vitamin C	Citrus Fruits, Amla, Vegetables and Liver
Vitamin D	Mushrooms, Liver Eggs and Fish
Vitamin E	Green Vegetables, Variety of Fruits
Vitamin K	Spinach, Broccoli, and other green vegetables, Egg Yolk

5.7.1 Determination of Vitamin C (Ascorbic Acid)

INTRODUCTION

Vitamin C is a highly water-soluble compound that has both acidic and strong reducing properties. It naturally occurs in many plants and animals except in humans. The natural vitamin exists in L-ascorbic acid form. The D-isomer (i.e., D-ascorbic acid), which is the mirror image of the same molecular structure, has only about 10% of the activity of the L-isomer. L-ascorbic acid is a weak sugar acid which is structurally related to glucose attached to a hydrogen ion. It is a strong reducing agent, which carries out its reducing function and easily converts to its oxidized form, the L-dehydroascorbic acid, when oxidative stress is present. Due to this characteristic, L-ascorbic acid is commonly applied in food industry as a food additive functioning as a versatile antioxidant to protect foods from deterioration by oxidation.

Vitamin C is an essential nutrient in humans as it functions as a cofactor in several vital enzymatic reactions. It is widely known that deficiency of vitamin C would lead to scurvy in humans. Vitamin C also has other beneficial effects to our body, such as preventing common cold/heart diseases and strengthening human immune system. However, human beings cannot synthesise Vitamin C by themselves and should obtain it from other sources. The richest natural sources of vitamin C are fruits and vegetables, for example, blackcurrant, blueberry, orange, lime, lemon, strawberry, cabbage and malt. It is noted that vitamin C can be chemically decomposed under certain conditions, such as heating and oxidation, many of which may occur during the cooking of food.

In general, the recommendation for vitamin C intake in humans is around 60–95 milligrams per day and the maximum upper intake level is 2000 milligrams per day.

In this experiment, we use titration method to determine the concentration of vitamin C in freshly prepared and packaged fruit juice samples.

Ionate ions are reduced to form iodine

$$IO_3^- + 6H^+ + 5e^- \longrightarrow 1/2\ I_2 + 3H_2O$$

While the iodide ions are oxidized to form iodine

$$2I^- \longrightarrow I_2 + 2e^-$$

Combining these half equations demonstrates the reaction between iodate and iodide ions

$$IO_3^- + 5I^- + 6H^+ \longrightarrow 3I_2 + 3H_2O$$

It is the iodine formed by this reaction that oxidizes the ascorbic acid to dehydroascrobic acid as the iodine is reduced to iodide ions.

$$\text{Ascorbic acid} + I_2 \longrightarrow 2I^- + \text{dehydroascrobic acid}$$

MATERIALS REQUIRED

Apparatus: Hot Plate, Timer, Glass Thermometer, Micropipette and Pipette Tips, Titration Set (Burette, Stand, Clamp, Tile and funnel).

Glassware: Conical flask, Buchner funnel, Filter paper, Glass stopper bottles, Measuring cylinder, Beaker, Test tube test tube rack.

Reagents:

1. Ascorbic acid standard solution (100 mg L-ascorbic acid in 100 ml of 0.5% (w/v) Oxalic acid)
2. 0.5% (w/v) Oxalic acid (extracting media)
3. 2,6-Dichloro-Indophenol (DCIP) Solution
4. 30% Hydrogen peroxide

Sample/s Used: Fruit Juice/Beverage Samples

PROCEDURE

I. Standardization of DCIP (Titrant) Solution

1. Using a 25 ml measuring cylinder, measure 25 ml of 0.5% oxalic acid and transfer it into a 250 ml conical flask.
2. Using a micropipette, add 1 ml of ascorbic acid standard solution into a 250 ml conical flask and mix it well with the oxalic acid.
3. Carry out a trial run of titration. Titrate the ascorbic acid solution rapidly with the DCIP solution. Add the DCIP solution through the burette and vortex the solution well. You may observe the DCIP solution changes to pink color when contacts with the ascorbic acid solution and then becomes colorless after shaking well. The end point is reached once a drop of DCIP solution is added and a distinct light rose-pink color persists in the solution even after mixing thoroughly. You may add the DCIP solution in excess as in this trial run, you are only required to be familiar with the color of the end point and to estimate the volume of DCIP solution required to bring the vitamin C solution to the end point.
4. After the trial run, conduct another three actual titrations to the ascorbic acid standard solution and average the results. At this time, add the DCIP solution drop by drop carefully when the volume of DCIP solution added is close to the end point volume. (You may prepare three ascorbic acid solutions at the same time with three conical flasks.)
5. Record the volume of DCIP solution added. Estimate the reading of the burette up to two decimal places (e.g., 6.15

ml or 6.20 ml, but not 6.2 ml). This is equal to the volume of DCIP solution required to completely react with 1mg of ascorbic acid.

II. Determination of the Vitamin C concentration in a fresh lemon juice

1. Cut a lemon in half with knife and squeeze the juice out.
2. With the aid of a Buchner funnel and filter paper, separate the flesh and seeds from the juice. (~10 ml lemon juice should be collected)
3. Pipette 1 ml of the lemon juice into a 250 ml conical flask, which contains 25 ml of 0.5% oxalic acid, and mix the solution well.
4. Titrate the lemon juice solution with the DCIP solution.
5. Triplicate the test and average the results.
6. Calculate the vitamin C concentration in the fresh lemon juice by the following equation,

$$\text{Conc. in juice (mg/ml)} = \frac{\text{DCIP (ml) used to titrate lemon juice}}{\text{DCIP (ml) used to titrate standard}} \times 1\ \text{mg/ml}$$

** assuming that the amount of titrant for the blank (1 ml H_2O + 25 ml of 0.5% Oxalic acid) is very little and insignificant.

III. Determination of the vitamin C concentration of an apple juice beverage and investigation of its heat stability

1. Open the package of the beverage (apple juice) and transfer ~16 ml into a test tube (16 × 150 mm).
2. Pipette 1 ml of the beverage into 25 ml of 0.5% oxalic acid in a conical flask.
3. Titrate the sample with the DCIP solution using the procedures previously mentioned. Obtain 3 titration readings in total and average the results.
4. Put ~200 ml water into a 250 ml beaker. Put it on the hot plate and turn on the heater. Let the water heat up to boil.
5. Put the test tube containing the apple juice sample into the boiling water bath. Put a thermometer into the test tube and let the juice sample heat up to ~100°C.
6. When the sample temperature has reached to 100°C, start to count the time. Collect the apple juice sample at 15 min, 30 min and 45 min intervals. Conduct the DCIP titration thrice (3 times) each similarly as above.
7. Compare the vitamin C concentration of the apple beverage at different time intervals after boiling with the unboiled one.

RESULTS

I. Stanardization of DCIP (Titrant) solution

	Initial reading	*Final reading*	*Volume of DCIP (ml) used to titrate with 1 ml of the ascorbic Acid standard solution (1 mg/ml)*
		Average	

II. Vitamin C concentration in the fresh lemon juice

	Initial reading	*Final reading*	*Volume of DCIP (ml) used to titrate with 1 ml of lemon juice*
		Average	

The vitamin C concentration of the fresh lemon juice is _____ (mg/ml).

III. Vitamin C concentrations of the apple juice with different heatinng time

	Volume of DCIP (ml) used to titrate with 1 ml of apple juice when			
	Without heating	*Heating (100°C) for 15 min*	*Heating (100°C) for 30 min*	*Heating (100°C) for 45 min*
Average				

The vitamin C concentration of the apple juice without heating is (mg/ml).

Vitamin C chemically decomposes under certain conditions, many of which may occur during the cooking of food. Vitamin C concentrations in various food substances decrease with time in proportion to the temperature they are stored at. Cooking can reduce the vitamin C content of vegetables by around 60% possibly partly due to increased enzymatic destruction as it may be more significant at sub-boiling temperatures.

PRECAUTIONS

1. Careful handling of the hot plate and hot-water bath is required they may cause severe burn.
2. Pay attention when handling the hot glassware.
3. Wear protective glove when handling the test tube in hot water.

REVIEW QUESTIONS

1. What is the effect of heat on vitamin C?
2. Does ascorbic acid plays a role in enzymatic browning?
3. Vitamin C content as determined by the 2,6-dichloroindophenol method be underestimated or overestimated in the case of heat processed juice samples. Why?

5.8 MISCELLANEOUS

5.8.1 Study of Sol-gel Properties of Different Starch Samples

INTRODUCTION

Starch is the principal food-reserve polysaccharide of the plant kingdom, where it occurs in cereal grains, pulses, tubers, bulbs and fruits in varying amounts ranging from a few percent to over 75% in the case of grains. Starch provides the main source of energy in the diet of a man. Refined starches, either natural or modified cereal starches usually contain 25% amylose and 75% amylopectin. It is the structure of amylose that is responsible for the blue color produced by iodine with starches and also contributes to the gelling characteristics of cooked and cooled starches. Amylopectin is responsible for the thickening properties of starch preparations but it doesn't contribute to gel formation.

If a mixture of starch and water is heated, the intra-molecular hydrogen bonding is broken, the grains absorb more water and swell. The swelling causes loss of birefringence. The temperature at which the granules begin to swell rapidly and lose birefringence is called "gelatinization temperature". This change is completed with a rise of few degrees in temperature after gelatinization sets in. Thus, it is more appropriate to call the temperature of gelation, "the Gelatinization temperature Range".

The course of gelatinization depends not only on botanical origin of starch and temperature used but also on the water content of suspension. Thus, dried starch with 1-3% water undergoes only slight changes upto a temperature of 180°C whereas starch with 60% water completely gelatinizes at as low as 70°C.

MATERIALS REQUIRED

Apparatus: 250 ml beakers, Glass rods, Weighing balance, Heat source.

Sample/s Used: Corn flour, Arrowrot, Maida, Besan, Wheat flour.

Additives Used: Common salt, Sugar, Oil and Citric acid.

PROCEDURE

I. Determination of Gelatinization Temperature Range

1. 10% starch solution (100 ml) was prepared for all given samples in separate beakers.
2. The mixture was heated on flame and the consistency was noted at different temperatures.

3. The temperature at which the mixture starts leaving the sides of beaker is noted as initial gelatinization temperature.
4. The temperature at which the mixture thickened was noted as the final gelatinization temperature.

II. Determination of effect of concentration on GTR

1. 20% solution of sample was made and above process was repeated.

III. Determination of effect of additives on GTR

1. 10% starch solution was prepared and given additive @1% was added and the effect of GTR studied. The procedure was repeated for each given additive.

OBSERVATIONS

Table showing determination of GTR and influence of concentration on GTR

Sample	*Conc. (%)*	*Initial GT (gelatinization temperature) (°C)*	*Final GT (gelatinization temperature) (°C)*	*GTR*

Table showing effect of additives on GTR

S. No.	*Sample*	*Additive used*	*Gelatinzation temperature (°C)*	
			Initial	*Final*

RESULTS

GTR is a range which begins at basic level of gelation and ends at a point at which entire contents become gel like in consistency (thick and translucent). The effect of concentration shows that higher is the conc., faster is the gelatinization process. Effect of additives like salt, sugar, acid and oil was also studied and it was found that all three additives except acid increase the GTR.

In case of salt, which is an electrolyte, Na^+ and Cl^- are formed which interact with H^+ and OH^- of water, thus making water unable to enter the starch granules. As a result of this, gelatinization temperature increases on adding salt.

Sugar is hygroscopic and takes away water. It limits the swelling of starch granules while competing with water and increases the gelatinization temperature.

Oil forms a coating on starch granules and therefore prevents the entry of water into starch granules as a result gelatinization temperature is increased.

On the other hand, when acid is used, especially a weak acid like citric acid, it modifies the structure of starch granules because of the polarity that the starch-citric acid complex acquires. As a result, a decrease in gelatinization temperature is observed.

Temperature does affect gelatinization process. At initial gelatinization temperature, solution gets thickened. Higher the concentration, faster is the gelatinization as more starch granules increase the probability of water molecules entering the starch granules. Therefore, increasing the concentration from 10% to 20% decreases the GTR. Although, effect of additives on GTR is variable depending upon the fact, whether they are making the water available or unavailable to the starch granules.

PRECAUTIONS

1. Thermometer should not touch the base of beaker.
2. GTR should be properly noted.
3. Additives should be thoroughly mixed.

5.8.2 Determination of the Thermal Inactivation Time of Different Enzymes Present in Fruits and Vegetables

INTRODUCTION

Enzymes are biological catalysts that are present in minor amounts in many foods, but play a major role in foods. Most of the enzymes remain active even after the plant is harvested or an animal is killed and promote a wide variety of biochemical reactions. Enzymes may change the composition of foods; in some cases, such changes are desirable but in most instances are undesirable, so the enzymes must be deactivated.

It is very important to know the time and temperature combination at which an enzyme is inactivated, for the correct preservation of a food, e.g., in canning industry. Moreover, this also helps to save the fuel and expenditure of the processor as further heat treatment of the food, beyond the inactivation temperature is not required. Enzymes are used as indicators in analytical methods, e.g., inactivation of peroxidase enzyme acts as an indicator in the blanching of fruits and vegetables. The thermal degradation time of an enzyme varies with the variety of enzyme, the percentage of enzyme present in the food, composition of the food, genetic variation, etc.

The enzymes considered in the experiment are:

1. **Catalase:** It catalyses the conversion of two molecules of hydrogen peroxide into water and molecular oxygen as follows:

$$2H_2O_2 \xrightarrow{\text{catalase}} 2H_2O + O_2 \uparrow$$

2. **Peroxidase:** The reaction catalyzed by peroxidase involves hydrogen peroxide as an acceptor and a compound of AH_2 as a donor of hydrogen atoms (e.g., guaiachol):

$$H_2O_2 + AH_2 \xrightarrow{\text{peroxidase}} 2H_2O + A$$

   In contrast to the reaction of catalase, no molecular oxygen is formed. This enzyme is quite heat stable and retains about 5% of its activity even after heating at 85°C. Peroxidase leads to off flavours in vegetables and contributes to browning action in some fruits and vegetables. The peroxidase test is used as an indicator of satisfactory blanching of fruits and vegetables.

3. **Polyphenoloxidase:** It is a group of enzymes which are involved in enzymatic browning. This group includes enzymes like phenol oxidase, cresolase, tyrosinase, catecholase, etc. These enzymes have the capacity to oxidize phenolic compounds to o-quinones:

The action of polyphenolases is detrimental when it leads to browning in fruits and vegetables but is beneficial in the processing of tea and coffee. Relatively high levels are found in potatoes, mushrooms, apples, peaches, bananas, etc. This enzyme also catalyses the conversion of monophenol into o-diphenols.

Reactions Involved

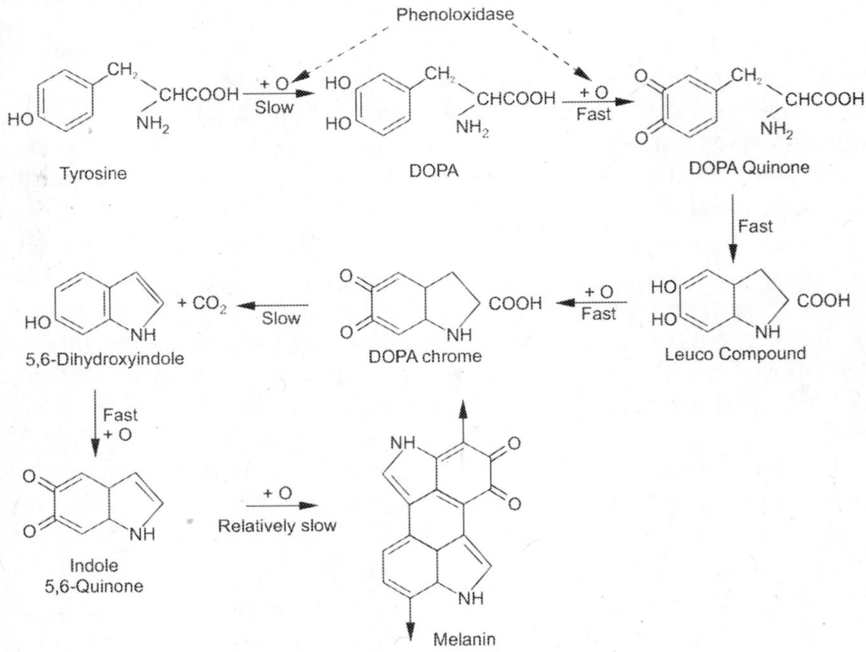

MATERIALS REQUIRED

Apparatus: Pan, Sieve, Heat source, Pestle and mortar,

Glassware: Stoppered test tubes

Reagents:

1. 10% H_2O_2 in water
2. 1% Guaiachol solution in absolute alcohol

Sample/s Used: Fresh vegetables such as potato, ridge gourd, bottle gourd, beans and fruits such as apple.

PROCEDURE

I. Preparation of Enzyme Extract

1. Take the given sample. Wash, peel and slice in chilled water.
2. Weigh 10 gm each.
3. Dip in boiling water as per requirement, i.e. 0, ½, 1, 1 ½, 2, 3, 4, 5, … 10 minutes…(till enzyme inactivates).
4. Dip in cool water (30 ml) after removing from boiling water bath and crush in a pestle and mortar.

5. Filter through muslin cloth or sieve. This is ENZYME EXTRACT and dispense 5 ml each in three separate test tubes.

II. Checking the Presence of Enzymes

1. **Catalase test:** Add 1 ml of 10% to 5 ml of enzyme extract. Note effervescence for positive test.
2. **Phenolase test:** Leave 5 ml enzyme extract in test tube as such. If contents turn brown within 3 minutes, it confirms the presence of phenolase.
3. **Peroxidase test:** To the 5 ml enzyme, extract add 1 ml of guaiacol followed by 1 ml of 10% H_2O_2 along the sides of test tube. Formation of brown ring indicates presence of peroxidase enzyme.

   Repeat the above steps each time you perform thermal treatment for varying lengths of time.

OBSERVATIONS

Table showing thermal inactivation time for enzymes present in the given samples

S. No.	*Sample*	*Enzyme tested*	*Thermal inactivation time (in min)*					
			0	*1/2*	*1*	*2*	*3*	

RESULTS

The thermal inactivation time for different enzymes present in various samples are shown below:

S. No.	*Sample*	*Enzyme*	*Thermal inactivation time (min)*

Two of the more heat-resistant and widely distributed enzymes in plant tissues are peroxidase and catalase. Activity of these enzymes, therefore, can be used to evaluate the effectiveness of a blanching treatment. If both are inactivated then it can be assumed that other significant enzymes also are inactivated. The heating time necessary to destroy catalase or peroxidase depends on the type of fruit or vegetable, the method of heating, the size of the fruit or vegetable, and the temperature of the heating medium. Heating media other than water (e.g., steam, hot air and microwave) and at temperatures other than 100°C can be used. Although blanching of vegetables is most often done in hot water or steam, blanching of fruits is often done in calcium brines to firm the fruit through the formation of calcium pectates.

PRECAUTIONS

1. The sample should be kept immersed in cold water to prevent browning.
2. Guaiacol should be added along the sides.
3. H_2O_2 and guaiacol should be stored in dark bottles.
4. Crushing and grinding of fruits and vegetables should be done in minimum time.
5. 1% solution of guaiacol should be made in absolute alcohol.
6. Control should not be dipped in boiling water.

5.8.3 Estimation of the Extent of Non-enzymatic Browning

INTRODUCTION

The non-enzymatic browning reactions are responsible for the color and flavour of foods, such as dates, honey and chocolate. The distinctive flavour that coffee beans, groundnuts, cashew nuts and breakfast cereals develop after roasting is also due to browning reactions. The presence of reactive reducing sugars is responsible for browning in foods. On heating they undergo "ring-opening, enolisation, dehydration and fragmentation". The unsaturated carbonyl compounds that are formed react to produce brown polymers and flavour compounds. Since, there is no role of enzymes in the formation of these brown to black polymers, the reactions are called "non-enzymatic browning". These include caramelization, the Malliard reactions and ascorbic acid oxidation. Dehydration and decarboxylation can lead to the formation of furfural, which can polymerise to form brown pigments or combine with amino acids in the strecker degradation.

MAILLARD REACTION

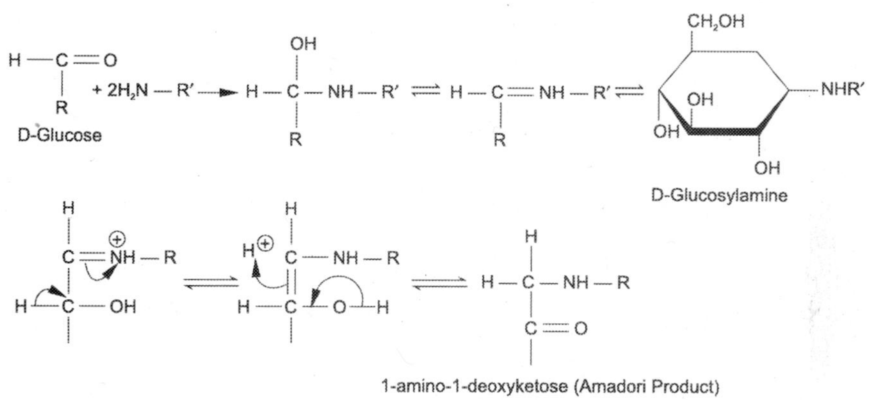

The ketosamine products can either dehydrate into reductones and dehydro reductones, which are caramel, or produce short chain

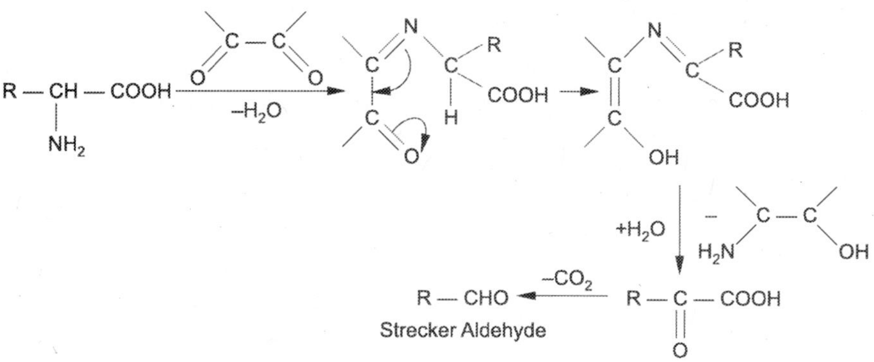

hydrolytic fission products such as diacetyl, acetol or pyruvaldehyde which then undergo the Strecker degradation.

PRINCIPLE INVOLVED

In the experiment, the sample which has undergone browning is taken and 60% alcohol is used to dissolve pigments. It is then analyzed for O.D. using spectrophotometer.

The quantitation of brown pigments in sample is based on Beer's law which states that amount of light absorbed is proportional to the concentration and path length, i.e., A = KCl. If path length is kept constant, concentration can be estimated.

MATERIALS REQUIRED

Apparatus: Spectrophotometer, Whatmann paper no. 4.

Glassware: Stoppered conical flask, Test tubes, Funnel, Cuvette.

Reagents: 60% Alcohol.

Sample/s Used: Besan, Semolina

PROCEDURE

1. Take 5 g of sample in a stoppered conical flask.
2. Then add 100 ml of 60% alcohol.
3. Mix the contents and keep overnight.
4. Filter the contents using whatmann no. 4.
5. Take the optical density at 440 nm for different samples and compare using 60% alcohol as blank.

OBSERVATIONS

S. No.	*Sample*	*Dilution factor (a)*	λ_{max} *O.D.(b)*	*Net O.D. (a × b)*

RESULTS

The O.D. is directly proportional to the extent of browning. Two samples (besan and roasted besan) were taken and the extent of non-enzymatic browning was estimated. The net O.D. (at 440 nm) for roasted besan was found out to be higher than that for unroasted besan. The higher value in the case of roasted besan was because it underwent non-enzymatic browning during roasting (prolonged heating).

PRECAUTIONS

1. Weighing should be accurate.
2. Sample should be completely soaked.
3. Filtration should be done using Whattman no. 4
4. O.D. should be taken at λ_{max} O.D.

REVIEW QUESTIONS

Gelatinization

1. What are modified starches?
2. What is retrogradation and gelatinization?
3. What is the effect of concentration on GTR range?
4. What is the sparkling white paste formed during gelatinization?
5. Give the application of GTR in the food industry.
6. What is the role of gelatinization in the cooking of noodles?

Browning

1. What is non-enzymatic browning and give its types?
2. What is enzymatic browning and name the enzymes involved in it?
3. What is the type of browning that occurs in citrus juices?
4. Name the type of enzymatic browning that requires the presence of oxygen.
5. What are the different methods to determine non-enzymatic browning?

Thermal Inactivation

1. What is the role of peroxidase, catalase and PPO in reference to the food products?
2. Name the enzyme which is the index of blanching.
3. Describe the role of the PPO enzymes in fruits.
4. Define blanching.
5. Which quality is affected by the action of the enzymes catalase and peroxidase in fruits and vegetables?
6. State the difference between the action of enzymes catalase and peroxidase.
7. Give another name for lipolytic enzymes.

6

Instrumental Techniques

Principles and working of some common laboratory equipment used the food analysis is described below.

6.1 BROOKFIELD VISCOMETER

6.1.1 Principle

Viscometers are used to measure viscosity of test fluids at a given shear rate. Viscosity is a measure of fluid's resistance to flow. Brookfield viscometer employs the principle of rotational viscometry. It measures the torque required to turn an object, such as a spindle, in a fluid which indicates the viscosity of the fluid. Viscosity is found through the following relationship:

$$\eta = \frac{\tau}{\dot{\gamma}}$$

where η is the viscosity in poise, τ is the shear stress in dynes/cm^2, and γ is the rate of shear (sec^{-1}). The magnitude of viscosity is influenced by temperature, type of spindle, time of test, spindle speed, container and other sample preparation procedures that may affect the behavior of the fluid.

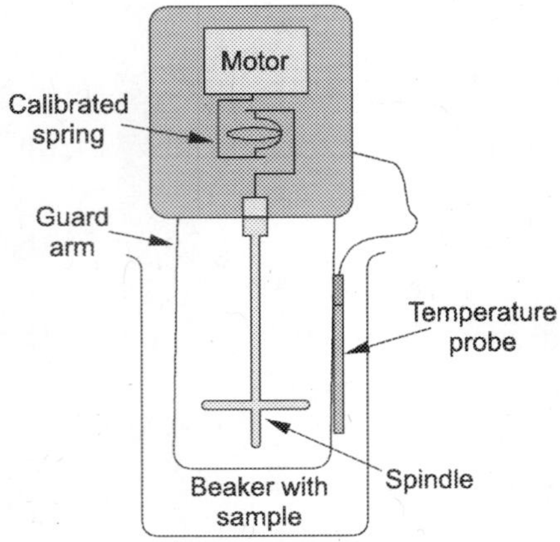

Working of a viscometer

6.1.2 Working

The spindle is inserted in the test material in a container until the fluid's level is at the immersion groove on the spindle's shaft. This spindle is driven by a motor through a calibrated beryllium-copper spring which applies the torque; deflection of the spring is indicated either by a pointer and dial or a digital display which measures the viscous drag of the fluid against the spindle. The amount of viscous drag is proportional to the amount of torque required to rotate the spindle and thus, to the viscosity of a Newtonian fluid. In the case of non-Newtonian fluids, Brookfield viscosities measured under the same conditions can be compared. Multiple speed transmission and interchangeable spindles are used to measure a variety of viscosity ranges. When developing a new test method, trial and error is often necessary in order to determine the proper spindle and speeds.

6.2 COLORIMETER

6.2.1 Principle

Colorimeters are instruments used for psychophysical analysis and provide measurements that correlate with human eye-brain perception. Colorimetric data is provided as tristimulus values. Three-dimensional scales, such as L*a*b* in close approximation to CIE system, have been developed to objectively quantify color values. This scale defines color as follows:

L* (lightness) axis: Black to white (0 to 100), white point at the origin

a* (red – green) axis: Positive values are red; negative values are green; 0 is neutral

b* (yellow – blue) axis: Positive values are yellow; negative values are blue; 0 is neutral

All visible colors can be quantified within this 3-D rectangular space. Hunter values may be used to calculate the total color difference (ΔE) between a sample and a standard in visual perceptibility units (National Bureau of Standards, NBS Units).

$$\Delta E = \sqrt{[\Delta L^2 + \Delta a^2 + \Delta b^2]}$$

6.2.2 Working

Three elements are necessary to see color: a light source, an object and an observer.

A light source emits light that appears white, but when diffracted by a prism, represents all wavelengths in the visible light spectrum (400-700 nm). Colorimeters use illuminants, which are plots of relative energy versus wavelengths, to represent different light sources under standardized and quantifiable conditions.

Objects or samples modify light differently, depending on the colorants (e.g., dyes, pigments) that are present, by selectively absorbing some wavelengths, while reflecting or transmitting others. The amount of light reflected or transmitted can be quantified to form a spectral curve of the object's color characteristics.

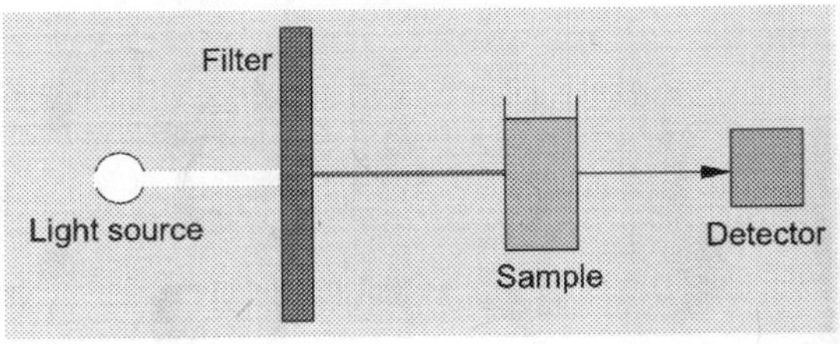

Working of a colorimeter

Finally, observer functions have been established using red, blue, and green glass filters, to simulate the cone-shaped receptors in the human eye and to quantify numerically the way in which the average human observer perceives color.

Example, in the rice industry, the values of greatest significance are L* (lightness) and b* (yellowness). Color differences may be expressed relative to a standard, such as a color tile, or may be expressed relative to other samples tested under uniform conditions. Sample uniformity is critical to the measurement. Readings may be affected by chalk, hulls or other interferences and therefore should be minimized when conducting color analysis.

6.3 pH METER

6.3.1 Principle

A pH meter provides a value as to how acidic or alkaline a liquid is. The basic principle of the pH meter is to measure the concentration of hydrogen ions. Acids dissolve in water forming positively charged hydrogen ions (H^+). The greater this concentration of hydrogen ions, the stronger the acid is. Similarly, alkali or bases dissolve in water forming negatively charged hydrogen ions (OH^-). The stronger a base is, the higher the concentration of negatively charged hydrogen ions there are.

The amount of these hydrogen ions present in a solution dissolved in some amount of water determines the pH.

6.3.2 Working

A pH meter consists of special measuring probe (a glass electrode) connected to an electric meter that measures and displays the pH reading. It is essentially a voltmeter with high input impedance which measures the voltage of an electrode sensitive to the hydrogen ion concentration, relative to another electrode which exhibits a constant voltage. The key feature of the pH-sensitive electrode is a thin glass membrane whose outside surface contacts the solution to be tested. The inside surface of the glass membrane is exposed to a constant concentration of hydrogen ions (0.1 M HCl).

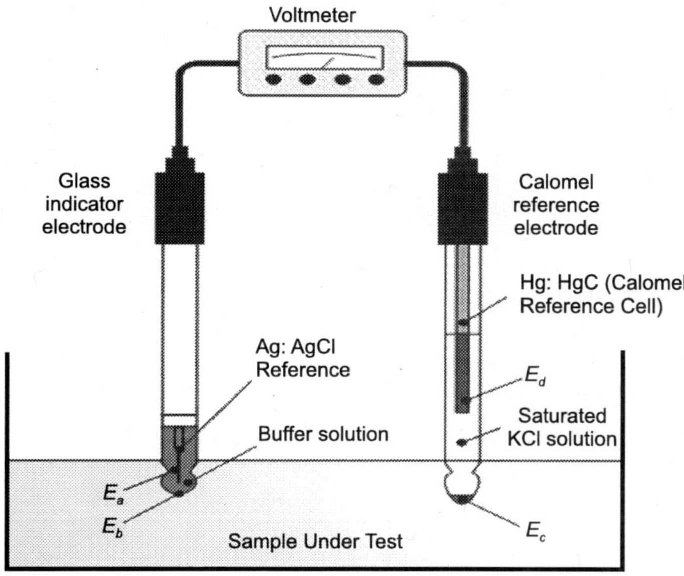

Working of a pH meter

When you place the probe or probes into a solution, the hydrogen ions in the solution will move toward the glass electrode replacing some of the metal ions on the glass coating. This will cause a tiny voltage across the glass bulb. This voltage is picked up by the silver wire and passed to the voltmeter where the voltmeter will amplify and change the voltage value into a value we can use, that is, pH units.

The greater the concentration of hydrogen ions, the greater the increase in voltage will be. This greater voltage corresponds to an increase in acidity causing the reading on pH meter to decrease. Similarly, a decrease in hydrogen ions or increase in hydroxide ions indicates an increase in alkalinity, causing a decrease in voltage and therefore an increase in the pH reading. To ensure accurate measurements from your pH meter, it needs to be calibrated both before first use and on an ongoing basis. This is usually done by dipping the probe into a buffer solution of a known pH and following the calibration steps based on our specific pH meter. It should be calibrated before each measurement; otherwise glass electrode will not give reproducible emf over long period of time. Electrodes should always be kept in distilled water when not in use.

6.4 ABBE'S REFRACTOMETER

6.4.1 Principle

Critical angle is the incident angle at which light - instead of getting to the other side of phase boundary - gets refracted in such a way that it becomes parallel to the phase boundary surface. For smaller incident angles, rays get through the boundary and, for larger they get reflected back. Critical angle can be easily calculated if we know refractive indices of both media. We start with the Snell's law:

$$\frac{\sin(\alpha)}{\sin(\beta)} = \frac{n_2}{n_{11}}$$

α is the incident angle, β is the refracted angle. When refracted light gets parallel to the surface, refraction angle is 90°. Construction of most refractometers is based on the fact that critical angle is very easy to detect - instead of measuring the angle at which light gets refracted, we just measure angle at which boundary between light and dark appears.

6.4.2 Working

The Abbe instrument is the most convenient and widely used refractometer. Figure 6.4 shows a schematic diagram of its optical system. The sample is contained as a thin layer (~0.1 mm) between two prisms. The upper prism is firmly mounted on a bearing that allows its rotation by means of the side arm shown in dotted lines. The lower prism is hinged to the upper to permit separation for cleaning and for introduction of the sample. The lower prism face is rough-ground: when light is reflected into the prism, this surface effectively becomes the source for an infinite number of rays that pass through the sample at all angels. The radiation is refracted at the interface of the sample and the smooth-ground face of the upper prism. After this, it passes into the fixed telescope. Two Amici prisms that can be rotated with respect to another serve to collect the divergent critical angle rays of different colors into a single white beam, that corresponds in path to that of the sodium *D* ray. The eyepiece of the telescope is provided with crosshairs: in making a measurement, the prism angle is changed until the light-dark interface just coincides with the crosshairs. The position of the prism is then established from the fixed scale (which is normally graduated in units of n_D). Thermosetting is accomplished by circulation of water through the jackets surrounding the prism.

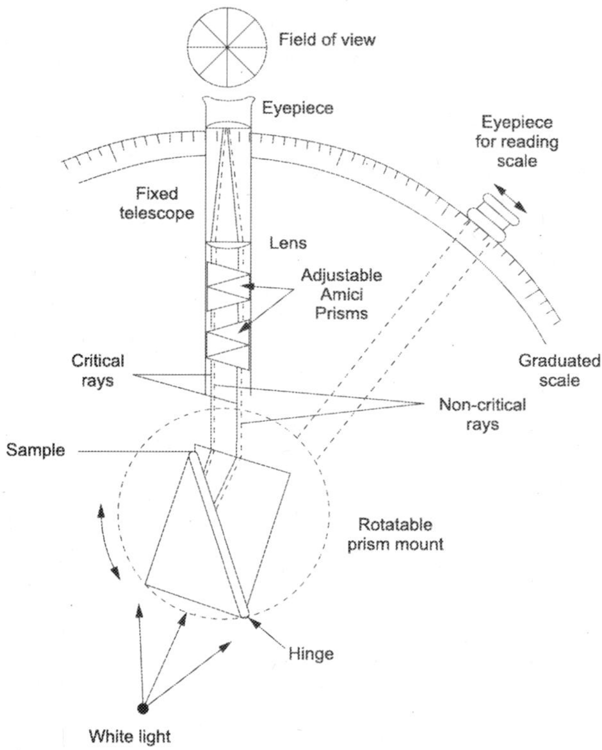

Schematic diagram of Abbe's refractometer optical system

6.5 SPECTROPHOTOMETER

6.5.1 Principle

Spectrophotometer works on the principle of Beer Lambert's law. When monochromatic light (light of a specific wavelength) passes through a solution, there is a quantitative relationship of the solute concentration and path length with the intensity of the transmitted light, that is,

$$I = I_o \times 10^{-kcl}$$

where I_o is the intensity of transmitted light using the pure solvent, I is the intensity of the transmitted light when the colored compound is added, c is concentration of the colored compound, l is the distance the light passes through the solution, and k is a constant. If the light path is a constant, as is the case with a spectrophotometer, Beer's law may be written,

$$I = I_o \times 10^{-kc}$$

$$T = \frac{I}{I_o} = 10^{-kc}$$

$$A = \log_{10} \frac{I_o}{I}$$

where k is a new constant, T is the transmittance and A is the absorbance of the solution. There is a logarithmic relationship between transmittance and the concentration of the colored compound.

The O.D. is directly proportional to the concentration of the colored compound. Most spectrophotometers have a scale that reads both in O.D. (absorbance) units, which is a logarithmic scale and in % transmittance, which is an arithmetic scale.

6.5.2 Working

A spectrophotometer consists of two instruments, namely a *spectrometer* for producing light of any selected color (wavelength), and a *photometer* for measuring the intensity of light. The instruments are arranged so that the liquid in a cuvette can be placed between the spectrometer beam and the photometer. The amount of light passing through the tube is measured by the photometer. The photometer delivers a voltage signal to a display device, normally a galvanometer. The signal changes as the amount of light absorbed by the liquid changes.

If the development of color is linked to the concentration of a substance in a solution, then that concentration can be measured by determining the extent of absorption of light at the appropriate wavelength. For example, hemoglobin appears red because the hemoglobin absorbs blue and green light rays much more effectively than red. The degree of absorbance of blue or green light is proportional to the concentration of hemoglobin.

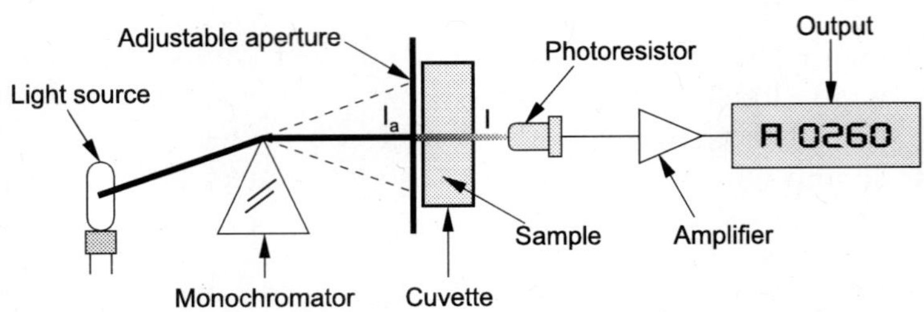

Working of a spectrophotometer

6.6 ELECTRONIC BALANCE

6.6.1 Principle

Balances are designed to meet the specific weighing requirements in the laboratory working environment. Electronic balance is a balance that generates current proportional to displacement of pan and is ideal for general purpose weighing applications.

There are two basic types of electronic balance designs:

1. Electromagnetic balancing type
2. Electrical resistance wire type (load cell type).

These are based on completely different principles, but what both have in common is that none of them directly measures mass. They measure the force that acts downward on the pan. This force is converted to an electrical signal and displayed on a digital display. As a means of measuring force, the electromagnetic balance method utilizes the electromagnetic force generated from a magnet and coil, whereas, the electrical resistance wire method utilizes the change in resistance value of a strain gauge attached to a piece of metal that bends in response to a force.

6.6.2 Working

6.6.2.1 Electromagnetic type

This is also called the "electromagnetic balance method." With mechanical balances, the sample is placed at one end of the beam and the weight is placed at the other end, and the value of the weight when both are perfectly balanced becomes the mass of the sample. With

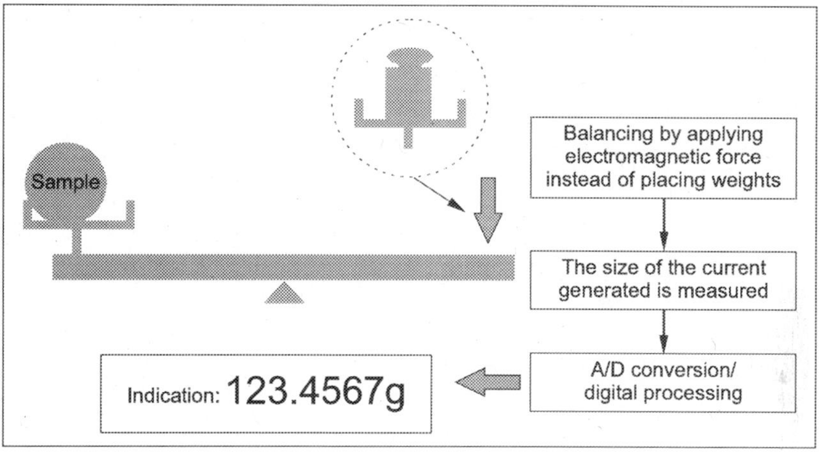

Working of an electromagnetic balance

electromagnetic type balances, an electrical force (electromagnetic force) is applied instead of a placed weight to balance the beam. The amount of electricity required for balancing the beam changes according to the weight of the sample that is placed. The amount of current when the beam is perfectly balanced is detected, and the mass is obtained from that detected value.

6.6.2.2 *Load cell type (electrical resistance wire method)*

One end of an object (elastic body) is made of aluminum which is fixed in place, and the sample is placed on the other end. The weight of the sample causes the elastic body to flex.

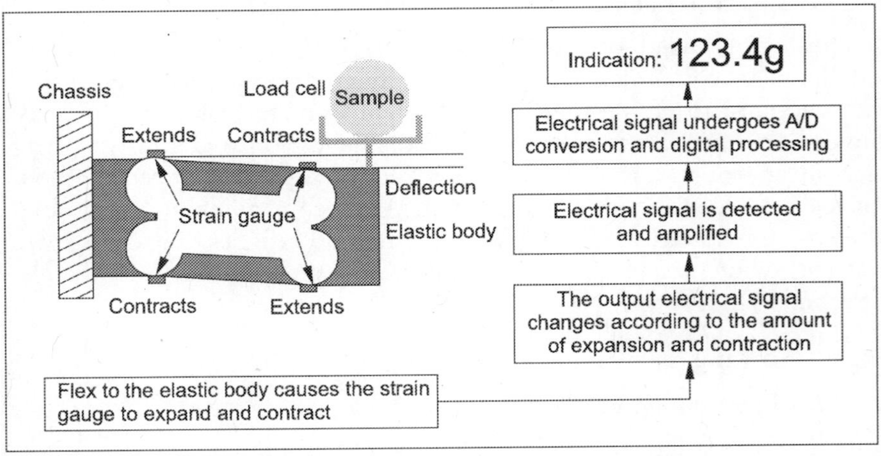

Working of a load cell type electronic balance

The amount of flex causes the strain gauges attached to the elastic body to expand and contract, changing the amount of electricity that is output (strictly speaking, the resistance value). The mass is then obtained from that amount of electricity.

Table 6.1 Comparison between electromagnetic type and load cell type

	Electromagnetic type	*Load cell type*
Advantages	High accuracy	• Simple structure • Even large models are easy to make
Disadvantages	Complex structure Difficult to downsize	• Accuracy is limited
Applications	Ultra-precision balances such as analytical balances	• Small, cheap balances that require only moderate accuracy • Large balances

6.7 VACUUM OVEN

6.7.1 Principle

Drying under vacuum is a widely used thermal process in many branches. The objective of this process is to gently dry high quality products containing water or solvents without modifying the product characteristics. During drying, the overall pressure in the drying oven is reduced so that water or solvents start changing from liquid to the vapor phase at low temperatures. Moisture in a weighed sample is removed by heating in an oven under specified conditions of time, temperature and vacuum and the residual dry substance weighed. The weight loss is calculated as moisture.

6.7.2 Working

In a vacuum oven, the thermal process occur in an airtight chamber where a desired level of vacuum has been applied using a vacuum pump. With this oven, the entire application can take place under vacuum or a controlled atmosphere can be created through the introduction of an inert gas. Vacuum oven speed up the out gassing process and is highly suitable for volatile substances.

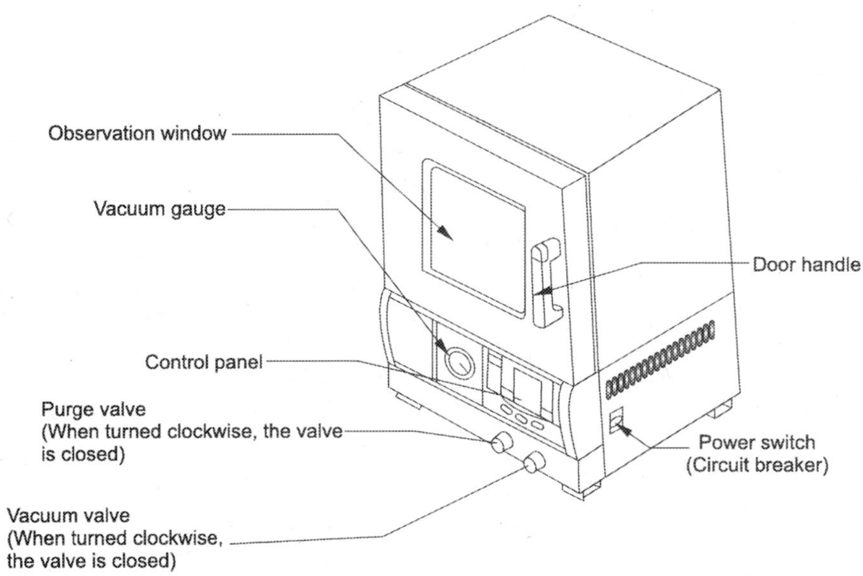

Schematic diagram of a vacuum oven

7

Related Terms

A. Methods of Expressing Concentration of a Solution

1. Physical Units
   (i) By expressing mass of a solute per unit volume of the solution. For example: 49 gm of H_2SO_4 per liter of solution.
   (ii) By expressing percentage of the solute in the solutions, i.e., the mass of the solute in 100 gm of the solution. For example: a 20% aqueous solution of KCl contains 20 g KCl in 100 gm of a solution.
2. Chemical Units
   (i) Molarity: It is the number of moles of solute present per liter of solution. For example: 1 M solution of NaOH contain 1 mole or 40 gm of NaOH per liter of solution.

$$\text{Molarity (M)} = \frac{\text{No. of moles of solute}}{\text{Vol of solution in litres}}$$

No. of millimoles (m mol.) = Vol. in ml × Molarity

   (ii) Normality: It is the number of gram equivalents of solute dissolved per liter of the solution.

$$\text{Normality (N)} = \frac{\text{Weight of solute} \times 1000}{\text{Eq. weight of solution} \times \text{Vol. of soln. in ml}}$$

$$\text{Equivalent weight} = \frac{\text{Molecular mass}}{\text{Acidity/basicity}} = \frac{\text{Atomic mass}}{\text{Valency}}$$

   (iii) Molality: It is defined as the number of moles of solute dissolved per 1000 gm (1 kg) of solvent.

$$\text{Molality (m)} = \frac{\text{Weight of solute in grams} \times 1000}{\text{Mol. weight} \times \text{Mass of solvent in grams}}$$

   (iv) Formality: It is the number of gram formula weights of an ionic substance dissolved per liter of solution.

$$\text{Formality (F)} = \frac{\text{Weight of substance} \times 1000}{\text{Grams formula wt.} \times \text{Vol. of soln. in ml}}$$

(v) Mole Fraction: It is defined as the ratio of the number of moles of a component to the total number of moles of all the components in the solution. For a binary solution, it is given as

$$XA = \frac{nA}{nA + nB}.$$

$$\text{where } nA = \frac{\text{Wt. of } A}{\text{Mol wt. of } A} \text{ and } nB = \frac{\text{Wt. of } B}{\text{Mol wt. of } B}$$

(vi) Parts per Million: It is defined as the number of parts by weight of a component dissolved in one million parts by weight of solution.

$$\text{ppm (A)} = \frac{\text{Weight of } A}{\text{Weight of solution}} \times 10^6$$

(vii) pH: It means 'puissance de hydrogen' which stands for power of hydrogen. It is used for expressing the concentration of H^+ ions or indirectly, OH^- ions.

The pH of a solution is defined as the negative logarithm of the H_3O^+ ion concentration in moles per liter.

$$\text{pH} = \log \frac{1}{[H_3O^+]}.$$

$$\text{or pH} = -\log [H_3O^+]$$

B. Preparation of Dilutions and Solutions

1. Simple dilution (dilution factor method based on ratios)

A *simple dilution* is one in which a *unit volume* of a liquid material of interest is combined with an appropriate volume of a *solvent* liquid to achieve the desired concentration. The *dilution factor* is the total number of unit volumes in which your material will be dissolved. The diluted material must then be thoroughly mixed to achieve the true dilution. For example, a 1:5 dilution (verbalize as "1 to 5" dilution) entails combining 1 unit volume of ***solute*** (the material to be diluted) + 4 unit volumes of the ***solvent*** medium (hence, 1 + 4 = 5 = dilution factor). The dilution factor is frequently expressed using exponents: 1:5 would be 5e–1; 1:100 would be 10e–2, and so on.

Example 1 Frozen orange juice concentrate is usually diluted with 4 additional cans of cold water (the dilution solvent) giving a dilution factor of 5, i.e., the orange concentrate represents one unit volume to which you have added 4 more cans (same unit volumes) of water. So the orange concentrate is now distributed through 5 unit volumes. This would be called a 1:5 dilution, and the OJ is now 1/5 as concentrated as it was originally. So, in a simple dilution, *add one less unit volume of solvent* than the desired dilution factor value.

Example 2 Suppose you must prepare 400 ml of a disinfectant that requires 1:8 dilution from a concentrated stock solution with water. Divide the volume needed by the dilution factor (400 ml/8 = 50 ml) to determine the unit volume. The dilution is then done as 50 ml concentrated disinfectant + 350 ml water.

2. Making fixed volumes of specific concentrations from liquid reagents

$V_1\,C_1 = V_2\,C_2$ Method

Very often you will need to make a specific volume of known concentration from stock solutions, or perhaps due to limited availability of liquid materials (some chemicals are very expensive and are only sold and used in small quantities, e.g., micrograms), or to limit the amount of chemical waste. The formula below is a quick approach to calculating such dilutions where:

V = volume, C = concentration; in whatever units you are working.

(stock solution attributes) $V_1\,C_1 = V_2\,C_2$ (new solution attributes)

Example 3 Suppose you have 3 ml of a stock solution of 100 mg/ml ampicillin (= C_1) and you want to make 200 μl (= V_2) of solution having 25 mg/ml (= C_2). You need to know what volume (V_1) of the stock to use as part of the 200 μl total volume needed.

V_1 = the volume of stock you will start with. This is your unknown.

C_1 = 100 mg/ml in the stock solution

V_2 = total volume needed at the new concentration = 200 μl = 0.2 ml

C_2 = the new concentration = 25 mg/ml

By algebraic rearrangement:

$$V_1 = (V_2 \times C_2)/C_1$$

$$V_1 = (0.2 \text{ ml} \times 25 \text{ mg/ml})/100 \text{ mg/ml}$$

and after cancelling the units,

$$V_1 = 0.05 \text{ ml, or } 50\ \mu\text{l}$$

So, you would take 0.05 ml = 50 μl of stock solution and dilute it with 150 μl of solvent to get the 200 μl of 25 mg/ml solution needed. Remember that the amount of solvent used is based upon the final volume needed, so you have to subtract the starting volume from the final to calculate it.

3. Moles and molar solutions (Unit = M = moles/L)

Sometimes it may be more efficient to use **molarity** when calculating concentrations. A **mole** is defined as one gram molecular weight of an element or compound, and comprised of exactly 6.023×10^{23} atoms or molecules (this is called Avagadro's number). The mole is therefore a unit expressing the *amount of a chemical*. The *mass* (g) of one mole of an element is called its **molecular weight (MW)**. When working with compounds, the mass of one mole of the compound is called the **formula weight (FW)**. The distinction between MW and FW is not always simple, however, and the terms are routinely used interchangeably in practice. Formula (or molecular) weight is always given as part of the information on the label of a chemical bottle.

The number of moles in an arbitrary mass of a dry reagent can be calculated as:

Number of moles = weight (g)/molecular weight (g)

Molarity is the unit used to describe the number of moles of a chemical or compounds in one liter (L) of solution and is thus a unit of *concentration*. By this definition, a 1.0 Molar (1.0 M) solution is equivalent to one *formula weight* (FW = g/mole) of a compound dissolved in 1 liter (1.0 L) of solvent (usually water).

Example 4 To prepare one liter of a simple molar solution from a dry reagent.

Multiply the *formula weight* (or MW) by the desired molarity to determine how many grams of reagent to use:

Chemical FW = 194.3 g/mole; to make 0.15 M solution use

$$194.3 \text{ g/mole} \times 0.15 \text{ moles/L} = 29.145 \text{ g/L}$$

Example 5 To prepare a specific volume of a specific molar solution from a dry reagent.

A chemical has a FW of 180 g/mole and you need 25 ml (0.025 L) of 0.15 M (M = moles/L) solution. How many grams of the chemical must be dissolved in 25 ml water to make this solution?

⇒ grams/desired volume (L) = desired molarity (mole/L) × FW (g/mole) by algrebraic rearrangement,

⇒ grams = desired volume (L) × desired molarity (mole/L) × FW (g/mole)

⇒ grams = 0.025 L × 0.15 mole/L × 180 g/mole after cancelling the units,

⇒ grams = 0.675 g

So, you need 0.675 g/25 ml.

4. Percent Solutions (% = parts per hundred or grams/100 ml)

Many reagents are mixed as *percent concentrations* as weight per volume for dry reagent OR volume per volume for solutions. When working with a dry reagent it is mixed as *dry mass (g) per volume* and can be simply calculated as the *% concentration (expressed as a proportion or ratio) x volume needed = mass of reagent to use.*

Example 6 If you want to make 200 ml of 3% NaCl you would dissolve 0.03 g/ml × 200 ml = 6.0 g NaCl in 200 ml water.

When using **liquid reagents** the percent concentration is based upon *volume per volume,* and is similarly calculated as *% concentration × volume needed = volume of reagent to use.*

Example 7 If you want to make 2 L of 70% actone you would mix 0.70 ml/ml × 2000 ml = 1400 ml acetone with 600 ml water.

To convert from % solution to molarity, multiply the % solution by 10 to express the percent solution grams/L, then divide by the formula weight.

$$\text{Molarity} = \frac{(\text{Grams reagent}/100\ \text{mL}) \times 10}{\text{FW}}$$

Example 8 Convert a 6.5% solution of a chemical with FW = 325.6 to molarity,

[(6.5 g/100 ml) × 10]/325.6 g/mole = [65 g/L]/325.6 g/mole = 0.1996 M

To convert from molarity to percent solution, multiply the molarity by the FW and divide by 10:

$$\%\ \text{solution} = \frac{\text{Molarity} \times \text{FW}}{10}$$

Example 9 Convert a 0.0045 M solution of a chemical having FW 178.7 to percent solution:

[0.0045 moles/L × 178.7 g/mole]/10 = 0.08% solution

5. Concentrated stock solutions – using "X" units

Stock solutions of stable compounds are routinely maintained in labs as more concentrated solutions that can be diluted to working strength when used in typical applications. The usual working concentration is denoted as 1×. A solution 20 times more concentrated would be denoted as 20× and would require a 1:20 dilution to restore the typical working concentration.

Example 10 A 1× solution of a compound has a molar concentration of 0.05 M for its typical use in a lab procedure. A 20× stock would be prepared at a concentration of 20 × 0.05 M = 1.0 M. A 30× stock would be 30 × 0.05 M = 1.5 M.

6. Normality (N): Conversion to molarity

Normality = n × M where *n* = number of protons (H^+) in a molecule of the acid.

Example 11 In the formula for concentrated sulfuric (36 N H_2SO_4), there are two protons, so, its molarity= N/2. So, 36 N H_2SO_4 = 36/2 = 18 M.

C. Indicators

Table showing different indicators used in chemical analysis.

Indicator	*pH range*	*Color in acidic solution*	*Color in alkaline solution*	*Used in titration*
Methyl Orange	3.0 – 4.4	Pink	Yellow	Strong acid v/s strong/weak base
Methyl Red	4.4 – 6.3	Red	Yellow	Strong acid v/s weak base
Litmus	6.0 – 8.0	Red	Blue	Strong acid v/s strong base
Phenolphthalein	8.2 – 10.0	Colorless	Pink	Strong/weak acid v/s strong base

D. *Refractive index of distilled water at various temperatures

Temperature (°C)	*Abbe refractometer reading*	*Temperature (°C)*	*Abbe refractometer reading*
15	1.3335	22	1.3328
16	1.3334	23	1.3327
17	1.3333	24	1.3326
18	1.3332	25	1.3325
19	1.3331	26	1.3324
20	1.3330	28	1.3322
21	1.3329	30	1.3320

**Source: Handbook of Analysis and Quaility Control for Fruit and Vegetable Products by S. Ranganna 2nd edition, page 876*

Note: The refractometer should be checked for accuracy befor reading. Allow the water to circulate through the instrument for a few minutes. Note the temperature and the refractive index of the distilled water. Compare the reading obtained with the readings given above. If the variation is more than ± 0.002 from the readings given in the table, adjust the refractometer to give the correct reading.

E. *FSSA Specifications for Edible Oils

Type of Oil	*Refractometer reading at 40°C*	*Saponification value*	*Iodine value*	*Free fatty acid Max. %*
Coconut oil	34.0-35.5	Min 250	7.5-10	3.0
Cottonseed oil	55.6-60.2	190-198	98-112	1.0
Groundnut oil	54.0-57.1	188-196	85-99	3.0
Mahuas oil	49.5-52.7	187-196	58-70	0.25
Mustard (rape-seed) oil	58.0-60.5	168-177	96-110[a]	3.0
Olive oil	53.0-56.0	185-196	79-90	3.0
Poppy seed oil	60.0-64.0	186-194	133-143	3.0
Safflower oil	62.0-64.0	186-196	135-146	3.0
Sesame oil[b]	58.0-61.0	188-193	105-115	3.0
Sesame oil[c]	60.5-65.4	185-190	115-120	3.0
Soybean oil	61.7-69.5	189-195	120-141	3.0
Maiza (com) oil	56.7-62.5	187-195	103-128	1.25
Almond oil	54.0-57.0	186-195	90-109	–
Watermelon seed oil[4]	55.6-61.7	190-198	115-125	3.0

[a]Polybromide test be negative, [b]Sesame oil expressed from black, brown white or mixed seeds; [c]Sesame oil from white sesame seeds grown in Tripura, Assam and West Bengal; [d]Moisture and insoluble matter should not be more than 0.25%

**Source*: *Handbook of Analysis and Quaility Control for Fruit and Vegetable Products by S. Ranganna 2nd edition, page 213*

References

1. Nielsen, S.S. (1994). Introduction to the Chemical Analysis of Foods. Jones and Bartlett Publishers, Boston.
2. Pomeranz, Y. and Meloan, C.E. (1994). Food Analysis: Theory and Practice. Chapman and Hall, New York.
3. Gruenwedel, D.W. and Whitaker, J.R. (1984). Food Analysis: Principles and Techniques. Marcel Dekker, New York.
4. James, C.S. (1995). Analytical Chemistry of Foods. Blackie Academic and Professional, Glasgow, UK.
5. Ranganna, S. (1986). Handbook of Analysis and Quality Control for Fruit and Vegetable Products. Tata McGraw-Hill Education, India.
6. Fennema, Owen R. (1996). Food Chemistry. Marcel Dekker, New York.
7. Kirk, R.S. and Sawyer, R. (1997). Pearson's Composition and Analysis of Foods. Addison Wesley Longman Ltd., Harlow, England.
8. Procter, A. and Meullenet, J.F. (1998). Sampling and Sample Preparation. In: Food Analysis, 2nd Edition. Aspen Publication, Gaithersberg Maryland.
9. Adams, J.B. 1991. Review: Enzyme inactivation during heat processing of food-stuffs. *Int. J. Food Sci. Technol.*, 26: 1–20.
10. A.O.A.C 17th edn, 2000. Official method 981.11 Oils and Fats.
11. Hedge, J.E. and Hofreiter, B.T. (1962). In: Carbohydrate Chemistry, 17 (Eds. Whistler R.L. and Be Miller, J.N.), Academic Press, New York.
12. Maynard, A.J. (Ed.) (1970). Methods in Food Analysis, Academic Press, New York, p. 176.
13. Lane, J.H., and Eynon, L. (1923). Determination of reducing sugars by means of Fehling's solution with methylene blue as internal indicator. *J. Soc. Chem. Ind. Trans.* 32–36.
14. Somogyi, M. (1952). Notes on sugar estimation. *J. Biol. Chem.* 200: 245.
15. The Culinary Institute of America (1996). The New Professional Chef, 6th edition, John Wiley and Sons.
16. http://people.umass.edu/mcclemen/581Toppage.html
17. http://www.fssai.gov.in/